Climate Variation
and
Climate Change
in Ireland

Climate Variation and Climate Change in Ireland

Proceedings of a one-day conference held at UCD on Monday 24th January 1994

Edited by

John Feehan

Environmental Institute
University College Dublin

Printer:
Walsh Printer, Roscrea, Co. Tipperary, Ireland
Tel.: 0505 - 21322

ISBN 1-898473-10-2

Cover picture: Dawn in early summer over the deserted village of Slievemore on Achill Island.

Photograph: John Feehan

Contents

PREFACE

John Maunder
Department of Geography
University College Dublin

In association with my appointment to the Department of Geography at University College, Dublin as Visiting Professor from September 1993 to February 1994, I was asked by Professor Frank Convery, Director of the Environmental Institute, to organise a one-day conference at the University Industry Centre on the subject 'Climate Variations, and Climate Change : The Irish response'.

In this connection I had pleasure in inviting a number of experts from Ireland to discuss this subject and I was particularly pleased with their response. Although it was not possible to cover all of the relevant subjects relating to 'Climate, Climate Variations, and Climate Change: the Irish Response' in the one-day conference, I believe that a good coverage was achieved. In particular, the key components of the WMO World Climate Programme - namely data, monitoring, applications, services, research, impact studies, and policy implications, and the desirability of an Irish National Climate Programme/Committee were discussed, as well as specific subject areas such as economics, insurance, agriculture, climate and history, climate models, water resources, peatlands, energy, and buildings.

The reader will be aware of the tremendous interest, and in some cases serious concern, in many countries in regard to the changing climate, as well as the natural climate variations including the apparent increase in severe weather events during the last decade. Many countries have developed policies during the last few years with regard to both the risks involved and the opportunities provided by the variations in the climate, as well as the human-induced climate change which the majority of scientists predict will become extremely relevant in the next 20 to 50 years.

As a visitor to Ireland,* I am naturally not familiar with the details of the specific responses of Ireland – at the social, economic, scientific, and political levels – to the issue of climate variations, and in particular climate change, but all of the experts which I invited to participate in the conference had particular expertise. In this regard the importance of recent Irish publications on this subject were noted (Department of the Environment, 1993; McWilliams, 1993).

It was hoped that the conference would be of particular importance to decision-makers throughout Ireland, especially at the national level, and with this in mind I asked all participants to address two particular questions relating to the theme of the conference as it applied to their individual areas of expertise:

* first, what *is* the present Irish policy with regard to the climate resources of Ireland, the climate variations which have occurred and are continuing to occur in Ireland, and the climate changes that are likely to occur in Ireland in the 21st century as a result of enhanced greenhouse gas emissions;

* second, what *should* be the Irish policy with regard to the same questions?

I am sure that a careful reading of the Proceedings of the Conference (for which I am indebted to John Feehan of the Environmental Institute), will provide at least some of the answers to these questions and be of value to decision-makers throughout Ireland.

References

Department of the Environment, 1993. *Ireland: Climate Change - CO_2 Abatement Strategy,* Dublin, Stationery Office.

McWilliams, B.E. (Ed.), 1991. *Climate Change : studies on the implications for Ireland,* Department of the Environment, Dublin, published by the Stationery Office, 193pp.

---

* My background includes over 20 years with the New Zealand Meteorological Service, 2 years on the staff of the World Meteorological Organisation, about 12 years working in Universities in New Zealand, Canada and the United States, as well as visiting appointments at the Bureau of Meteorology in Australia, the Atmospheric Environment Service in Canada, and the Stockholm Environment Institute. I have also been closely involved in the activities of the World Meteorological Organisation for about 20 years, and am currently the President of the WMO Commission for Climatology.

POLICIES FOR CLIMATE VARIATION AND CLIMATE CHANGE: AN OVERVIEW

Frank J. Convery
Environmental Institute
University College Dublin

In Ireland, the weather is a source of constant comment, discussion and speculation. This interest is not surprising: for most of us who do not actually earn our living by farming, it is still only a generation or two since our forbears did so - and weather has a great bearing on crop yields and profitability for farmers. Furthermore, our weather is highly predictable in its unpredictability. We know that rain, wind, sunshine, showers, gusts will all occur within a relatively short time: we're just not so sure what will happen this afternoon. Climatic determinists argue that our economy, our psychology, our culture, even our religion, our drinking habits and our sense of humour are heavily influenced by weather. It is therefore no mean challenge to begin to assess what our policies are in dealing with climate variation and climate change, and what they should be.

This distinction between variation and change is important. The former refers to the 'normal' variability which one can expect from day to day, week to week, season to season, and year to year. The latter refers to a possible long-term change in climate, induced, in the case with which this volume is mainly concerned, by human activity.

It is important to make this distinction: it is easy to become seduced by the 'glamour' of climate change, with its images of ice caps melting, Sandymount as the Venice of Ireland, and the production of fine wines from Chateau Gorey in Wexford. But it is climate variation which makes life interesting in the short run. Experience tells us that the River Lee will flood, that Corkonians will continue to suffer inconvenience and damage, that the Shannon will continue to flood, that periodic droughts will put pressure on water supplies in the east (notably in Dublin). Short term variation may at times influence the course of history. In his paper on *Climate and Irish History*, John Tyrell notes that 'the landing in Bantry Bay by the French fleet in 1796 was decisively frustrated by a series of severe easterly storms that wrecked and dispersed the French fleet', and this may have been a turning point in Irish history.

Climate Variation

What are the policy lessons to be learnt in order to deal with variation? The policy process has a number of policy instruments which can be brought to bear. These include (1) the provision of information, (2) investment in infrastructure, (3) the provision and implementation of regulations, (4) the use of incentives (including prices) and (5) the undertaking of research and development.

1. Provision of information:

Short-term information

The Meteorological Service provides a steady flow of information predicting short term (daily, up to a week) weather; presumably individuals and companies, from fishermen to the houseperson with clothes on the line, adjust their behaviour accordingly.

Provision of information regarding medium-term implications

What I have in mind here is not longer-term predictions, but telling people what the implications of past trends are in terms of the probabilities for the future, given existing infrastructure and management practices.

Examples include:

Category	Sample Information
Residents along River Dodder	How many times can they expect to be 'flooded out' over the next 20 years?
Fishermen	How many 'fishing days' can they expect of have annually? - best and worst cases, and the likely frequency thereof.
Outdoor Festival	Probability (by month) of getting 3 'dry' days in a row.
Organisers	
Water Consumers	Number of days where shortage is likely over the next decade.

The point of developing such information and disseminating it is that it allows producers and consumers to adjust their behaviour to accommodate such variation. It also diminishes the logic (if not the political potency) of the case which is always made whenever there is weather related damage, to the effect that the taxpayer should compensate those adversely affected.

The economic question is the following: to what extent will cost-saving behaviour be induced by better information, and at what cost is the latter produced? The practical question is the following: what media should be used, who should be the disseminators, and how should they be financed?

2. Investment - development of infrastructure

Infrastructure can be created to mitigate or overcome the negative implications of climate variation. Some examples:

* Reservoir capacity can be expanded.
* Irrigation systems can be installed.
* Leaking mains can be repaired.
* Rivers can be widened and deepened.
* Coastal defences can be built.
* All weather tourism facilities can be provided.
* Silage making capacity, which is less weather dependent, can replace hay making.

In some situations, such as the marine environment, the 'technical fix' may not be feasible because we don't know enough about the dynamics of the coastal environment. Furthermore, these can only be implemented at a cost and these costs should not exceed the ensuing benefits; the costs are of two categories - private costs, borne by the individual or company benefiting from the investment, and external costs associated with the investment, which are borne by others.

Thus, if a coastal community builds groins into the sea to trap sand, which then slows the erosion of their coast, this is probably at the expense of another coastal community which would have received this sand had it not been intercepted, but now finds that erosion is accelerating as a consequence of its artificially induced sand deficit. If a river wetland is drained, this can affect the habitat of birds and other wildlife, with negative implications for hunting, the passive enjoyment of wildlife, and - where species are endangered - may even result in the elimination of a species, and thereby diminish the genetic endowment of us all. Generally, because there are typically many potential beneficiaries of intervention, there is a logic to having a public body take on the investment. There is less logic in economics to having the public purse carry the costs of such intervention, but there is compelling *political* logic to so doing; because the potential beneficiaries often do not carry either the private or the external costs in relation to the interventions which might be undertaken, there is an almost infinite demand for this 'free' good.

The pressure for additional investment can be diminished considerably by ensuring that the beneficiaries carry some or all of the costs. This will help bring demand into equilibrium with supply, and - if the beneficiaries are given some role in decisions concerning the scale, design and management of the intervention - it should also ensure cost-effectiveness in design and operation.

3. Regulation

Regulations - such as zoning - can be employed to influence what goes where. A few examples of the use of this policy instrument are noted below:

Situation	Action
Floodplains	Zone them for uses compatible with periodic flooding.
Damp in houses	Building regs can insist on appropriate structural standards.
Erosion due to overgrazing	Regulate the numbers grazing in the watershed.
Wet tourists	Require grant beneficiaries to provide all-weather facilities.

Regulation can be a crude instrument, as it typically forces a standardisation which may not be appropriate to all circumstances.

4. Incentives, including price and tax incentives

If farmers are given a tax deduction for investing in facilities which, *inter alia*, foster production systems which are relatively independent of the weather, this will encourage such investment.

If competitive, effectively functioning insurance systems are appropriately designed, they can foster a socially efficient response to weather variation. Such policies should have an element of cost sharing and a rise in premium following claims; if such 'clawbacks' are not provided, beneficiaries will not be encouraged to invest in any protective strategy, confident in the knowledge that insurance will cover their full costs.

Climate Change

The core hypothesis of the proponents of the idea that climate is changing is that the temperature of the globe is increasing. This is caused at least in part by the actions of humans and their animals, which result in a rise in the concentrations of greenhouse gasses in the atmosphere. The effects overall may prove to be malign, although there may be individual 'winners'. According to this hypothesis, we need to stabilise the emission of greenhouse gasses as soon as possible.

Climate change sceptics have emerged who dispute a number of key elements of the core hypothesis. In particular, it is argued that the measurement of temperature is still somewhat inexact - e.g. increasing urbanisation may give rise to local temperature rises, and such rises are then used (erroneously) to support the climate change hypothesis. A second objection to the hypothesis is that the projections are based on very sophisticated models, which have not been validated against real data; furthermore, the adaptability of the global climate system to perturbations may be understated, in that the equilibrating effects of increasing cloud cover etc. may prove to be much more significant than has heretofore been recognised.

Ireland comprises an interesting case study of the issues with which a small country must contend in addressing itself to the climate change challenge:

1. Measuring climate change - interpreting the past and predicting the future

In *Datasets for Climate Change Studies in the Climatological Archive of the Irish Meteorological Service*, Des Fitzgerald outlines the key features of the Irish data gathering system. He notes that there are 7 climatological stations with long-term data which go back tothe turn of the century. These are as follows:

Station	Comments
Valentia (Cahirciveen)	Since 1892. Hourly values; computer data only since 1940.
Phoenix Park	From 1829. New temperature-measuring technology introduced in 1940s makes time series analysis fraught.
Malin Head	Temperature and rainfall since 1870s.
Markree Castle	Continuous since 1875, but recent cessation of sunshine recording.
Roche's Point	Temperature and rainfall since 1870s.
Belmullet-Blacksod	Over 100 years, but a significant change of site.
Birr	Since 1872, but three significant changes of site.

His discussion exemplifies why time series analysis is so fraught: changes in site and in measurement technology all combine to make analysis of the past difficult, before one even addresses the future.

In *Climate Variability in Ireland: Are there Signals Amongst the Noise*, Stuart Daultrey attempts to work his way through these complexities and derive a long term trend (although he uses a different mix of stations to those featured by Fitzgerald). He concludes that:

> For Ireland as a whole, there has been a statistically significant rise in annual minimum temperature at the rate of 1.1°C per century, compared to a (statistically non-significant) fall in the annual

maximum temperatures of 0.5°C, giving a (statistically non-significant) fall in the annual daily temperature range. There has also been a statistically significant fall in annual sunshine at a rate of 1.0 hour per century.

He notes that there is quite a variation in regional 'performance', with the three western stations showing consistently large and statistically significant decreases in hours of bright sunshine, in all seasons except autumn, whereas the three eastern stations show no statistically significant change other than small decreases in winter sunshine at Clones and Dublin Airport.

The conventional climatological data of the Meteorological Service are complemented by specialist stations, as exemplified by the Mace Head station in Galway described by Aodhagán Roddy in *Baseline Climate Observation at Mace Head Atmospheric Research Station.* He describes the monitoring of biogeochemical cycles, the greenhouse gases (methane, surface ozone, nitrous oxide) and the main chlorofluorocarbon gases (CFCs), micropollutants, mercury and other heavy metals in the air, ultraviolet radiation etc..

In regard to predicting the future, Peter Lynch outlines the strengths and limitations of general circulation models (GCMs) as simulators of weather patterns, and as predictors of the future, in his contribution on *Climate Modelling.* He concludes that: 'the overall success of climate models in simulating the present climate of the atmosphere is impressive', but in regard to their capacity to anticipate the future, he notes that 'we can be confident that the dependability of model guidance will grow with their increasing sophistication, so that detailed regional climate impact predictions may be reliable for the future, perhaps shortly after the turn of the millennium'.

In *Climate Scenarios for Ireland,* John Sweeney reviews the findings and characteristics of the various GCMs developed. By applying the approach to Ireland, he concludes that the climate change in prospect as noted below by Brendan McWilliams is plausible, but notes that the Atlantic may act as a buffer, muting the magnitude of change.

2. Climate Change and the Economy.

In his essay on *Implications for Ireland of Climate Change - an Overview,* Brendan McWilliams synthesises the findings of a study published in 1991 by the Department of the Environment (*Climate Change - Studies on the Implications for Ireland*). The authors of the study took as given the scenario outlined by the Intergovernmental Panel on Climate Change (IPCC) whereby, by the year 2030, annual average temperature would have increased by 2°C, precipitation would increase by 5-10 per cent in winter, and decrease by the same amount in summer, and mean sea level would increase by 18 mm. The findings can be summarised as follows:

Sector	Main Conclusion
Agriculture	Grass and tillage crop yields > 20%; decline in input industries (feed and fertiliser); increase in range of horticultural crops.
Forestry	>Productivity where soil water not limiting, and < when vice-versa.
The Green Mantle	Woodlands and sand dunes to benefit; peatlands to suffer serious damage.
Hydrology	Flood peaks delayed and reduced; regular water shortages resulting from increased demand, and (in summer) - reduced supply.
Sea Level	176,000 hectares at risk from flooding; increasing storm frequency; increased erosion of 15-20% per year.
Fisheries	Negative for anadromous fish, good for lobster and shell fish.

In *Climate Change and Economic Policy*, John Fitz Gerald cites earlier work showing that a carbon tax of $20 per tonne of carbon - as proposed by the European Commission - would, if the proceeds were used to reduce the cost of social insurance, result in an increase in Gross National Product of 0.4 per cent. This is a very important finding, as it highlights the possibility of what is called 'win win' situation, whereby the environment and the economy both 'win' as a result of policy intervention. However, such a policy would adversely affect the peat industry, and poorer consumers, so that some mechanisms to ease the burden on such interests should be a part of policy.

The insurance industry is a significant short-term indicator of the sharp edge of climate, as damage accompanying such change shows up immediately in the form of increased claims, and subsequently in rising insurance premia. In *The Insurance Industry's View of Climate Variations*, A.F. Dlugolecki and J. Fraser note that, unlike the UK, the weather-induced claim experience in Ireland has not risen significantly in recent years. They go on to outline a number of measures which would allow a less traumatic transition to a world of higher storm and flooding frequency induced by climate change, including re-assessment of construction standards, the exemption of equalisation reserves, increase in the level of storm excesses, dialogue with re-insurers to ensure that Ireland is not 'lumped in' with countries less fortunate in regard to climate-induced damage, monitoring of scientific progress etc..

3. Climate Change and the Energy Sector

The greenhouse gasses which predominate are carbon dioxide (CO_2), methane (CH_4), nitrous oxide (N_2O) and chlorofluorocarbons (CFCs). The latter are by far the most potent, and are also the primary agents in the depletion of the ozone layer. As such, CFCs have become the focal point of much policy attention independent of greenhouse gas concerns.

CO_2 comes mainly from burning fossil fuels - coal, oil, natural gas, peat - and burning vegetation, mainly forest fires. Methane is produced from a variety of sources, including farming (flatulent animals, and waste degradation), wetlands, rice paddies, landfills, and gas drilling. An important source of nitrous oxide is nitrogenous fertiliser.

In the European Union, Ireland comes second after the Netherlands as a source of greenhouse gasses per capita. This is mainly because of our predominance as a livestock producer and therefore as a source of methane, the significance of peat as a source of fuel which is relatively high in carbon, and because we have no nuclear power, and very little hydro power (both of which are carbon-free), to mitigate the production of CO_2. In his paper on *Predicted Climate Change and Electricity Generation*, Owen Wilson highlights the measures which the ESB is undertaking to reduce the rate of growth in CO_2 emissions, but notes that at a recent conference in Holland, 'it was reported that measures taken to control non-CO_2 greenhouse gases, responsible for 50% of the predicted effect, could be taken at much lower costs than CO_2 stabilisation'.

Wind power, which produces no greenhouse gases, has emerged as a potential means of contributing to the improvement of Ireland's position in regard to greenhouse. In *Wind Energy in Ireland - A Survey of Relevant Issues,* Rick Watson notes that, to produce 20 per cent of Ireland's electricity production (installed capacity of 1,376 MW, slightly larger in capacity than Moneypoint) would require 2,752 wind turbines, spread over 44,000 hectares. There is one wind farm operating in Ireland, consisting of 21 wind turbines with a total capacity of 6.45 MW, located at Bellacorick, County Mayo. The capital costs were subvented by a 55 per cent grant via the European Union's *Valoren* programme. Two further projects are planned for Arigna, County Roscommon, and Castleisland, County Kerry under the *Thermie* programme.

Peatland acts as a carbon 'sink'; the peat deposit captures carbon as plants take in carbon via the photosynthetic process and then the organic matter accumulates rather than being broken down. In *Peatland, Carbon Cycling and Policy Issues,* Charles Shier estimates that, as peat is exploited for energy production - an average annual off-take of 180 tonnes of fuel peat for combustion purposes - this results in the release of about 45 tonnes of carbon per hectare per year. The total carbon release from peat exploitation in Ireland is estimated at 1.7 million tonnes per annum; this compares with an annual capture by peat of carbon amounting to 312,000 tonnes, comprising 18 per cent of the carbon released. Measures which would improve the carbon balance include the afforestation of cut-over areas, and the restoration of cutover areas to conditions of mire formation. Afforestation initially increases the rate of CO_2 emissions, as improved drainage allows oxidation, but as the trees grow, they sequester carbon, and thereby reduce its release to the atmosphere.

4. Towards a Policy Framework

Mary Moylan outlines the Irish policy position in *Climate Change: Current EU Policy and the Irish Contribution*. The UN Framework Convention on Climate Change was signed by over 150 States at the Earth Summit in Rio in June 1992. This convention was ratified by the European Union and came into effect at the end of March 1994. The Convention calls for a stabilisation of greenhouse gas concentrations in the atmosphere; in compliance with its obligations, the EU policy is to stabilise Union emissions of CO_2 at 1990 levels by the year 2000. A number of policies have already been agreed at EU level, including energy labelling, monitoring of CO_2 emissions and mitigating measures, by Member States; a proposal for a carbon tax has also been tabled. Ireland's proposals and initiatives in this regard are as follows:

* CO_2 emissions at a maximum of 10.7 million tonnes by the year 2000, an increase of 20 per cent over the current level.

* An afforestation programme at the rate of 30,000 hectares per annum, to sequester 0.8 million tonnes per annum by the end of the decade, thereby reducing the net increase of CO_2 to 11 per cent.

* The ESB is to have 75 MW of alternative energy capacity (including renewables and Combined Heat and Power (CHP) in place by 1997.

* Energy conservation measures include a Demand Side Management (DSM) programme by the ESB, to limit growth in consumption to 3 per cent per annum, new insulation standards for new buildings to achieve a reduction of 2 per cent from this source by the year 2000, improved public transport, and reduced congestion resulting therefrom, especially in the Dublin area.

* An interdepartmental group has been set up to monitor implementation of the national strategy, focusing initially on an inventory of greenhouse gas emissions. An Intergovernmental Panel on Climate Change is working with the European Commission to develop acceptable methodologies, with a view to having adequate reporting arrangements in place by April 1995, when the first meeting of the parties to the Convention takes place.

Issues

Our large numbers of livestock - 6.9 million cattle, 8.9 million sheep, 1.4 million pigs, 12 million poultry and 0.07 million horses - put us almost at the top of the greenhouse gas emissions league in per capita terms. We are therefore fortunate (in terms of the extent and cost of policy measures required) that the policy focus at EU level has been on carbon dioxide. But our peat-dependency, and our lack of nuclear power and paucity of renewables still makes it difficult to comply within the stabilisation targets without incurring significant costs. The following implications seem to follow from our present situation:

(i) There may be significant benefits accruing to Ireland from global warming, and there is in any event considerable uncertainly as to the extent to which the phenomenon is "real"; in such circumstances, it is politically difficult to demand substantial sacrifices to either deal with a problem which we may not have, or to mitigate a temperature rise which would yield net benefits overall.

(ii) In the context of European Union, will we be allowed a net increase of 11 per cent in CO_2 emissions? Since the Union as a whole has committed itself to stabilising emissions, any increase in Irish emissions can only be achieved at the expense of a commensurate reduction elsewhere. Such a zero sum game situation is not conducive to easy agreement, although the absolute amounts requested are so small in an EU context that it may be possible to accommodate the Irish concession.

(iii) Making sure that the estimates of emissions in the base year (1990) are accurate is crucially important, as it defines the possibilities for the future.

(iv) Irish policy depends to a significant extent on the carbon sequestering role of new forests. But some forests, e.g. those established on peat, may result in a net increase in carbon emissions in the short term. How is this to be accommodated in the relatively short-term context of the year 2000?

(v) We can simultaneously present ourselves as being at the forefront of EU environmental policy, while still gaining net increases in employment and GNP, by adopting and implementing the carbon tax proposals of the EU. We should seize this opportunity, making 'side payments' to the peat industry and to low-income households to compensate them for the negative income effects of such a policy.

(vi) We have been very slow to move effectively in regard to renewables, notably in regard to wind power. Denmark made the crucial investments in R&D and pilot project support over the past two decades, and has now captured important export markets. After an initial foray in the early seventies, we abandoned this sector completely; the inability to identify a sector in which we have potential comparative advantage, and to invest steadily in the necessary R&D and incubation of the entrepreneurs needed to carry it forward to commercialisation, goes a long way towards explaining why indigenous enterprise in Ireland still languishes.

We will need to substantially increase our renewables capacities above the target of 75 MW if we are to reduce the pressure to expand our fossil fuel capacity.

(vii) There seems to be little use of price signals to either producers (notably the ESB) or to consumers to conserve energy. Unless pricing sends an unambiguous signal that conservation pays, it is difficult to see how any of the other measures will have any great effect.

(viii) Work by Sue Scott (cited by John Fitz Gerald) seems to show very attractive energy conservation opportunities among existing households (i.e. measures which pay for themselves). Such opportunities win in financial, energy conservation and environmental terms. But there are few initiatives - in terms of better information, the use of prices, grant mechanisms etc. - designed to encourage the 'capturing' of these opportunities, although some are promised in the National Plan 1994-99.

(ix) There are many institutional and other interests fishing in the climate change pond. In: *The Role of the Irish Meteorological Service in Climate Change,* Declan Murphy outlines some of the cross-departmental and cross-disciplinary issues and needs. He outlines the potential of a National Climate Committee representative of the various public and private interests to stimulate research, to provide an authoritative voice on the topic in the country, to provide input into the development of strategies, and to inform the public and the media.

There is merit in having such a sounding board and promoter of research, which would also act as a synthesiser of information for the policy process and for the public. It would make a useful and probably cost-effective contribution if the following conditions were met: a clear brief from the relevant Ministers, especially Environment; an acceptance by the policy process that the Committee would be encouraged to advise on the formulation of policy; and a clear lifetime with specific, phased objectives. In the first instance, the Committee could advise the interdepartmental group which has been set up to monitor progress towards the achievement of CO_2 stabilisation targets (see point 4 above), with a view to completing phase one of its work by April 1995, when the EU monitoring system is to be installed.

CLIMATE, CLIMATE VARIATIONS, AND CLIMATE CHANGE: THE IRISH RESPONSE

John Maunder
Visiting Professor
Department of Geography
University College Dublin
(President WMO Commission for Climatology)

ABSTRACT
The Second World Climate Conference which was held in Geneva in November 1989 made a number of specific recommendations concerning the role nations need to play in the climate change dilemma. Understandably, most of these proposals were not particularly specific, nor were they written from the viewpoint of particular decision-makers in each of the 160+ countries of the world. Clearly much more specific guidance is needed by countries, and the realities of the marketing and trading world must also be appreciated and taken into account. This paper endeavours to look at the practicalities of climate change from a national point of view, and considers the importance of competitors - both now and in a warmer world. It also considers the urgent need to produce national econoclimatic impact studies - with both the present and the future climate taken into account, and the controversial problem of potential winners and losers amongst the 160+ countries of the world. The need to stress the continuing importance of the current as well as any future climate is highlighted, as well as the need to consider climate change as a challenge and not a threat. The various viewpoints on 'what is climate', 'what is a climate variation', and 'what is a climate change' are also examined.

I. Climate and Climate Change

In a recent paper on 'Global Warming: Change of Climate or Part of Climate' the author (Maunder 1993) re-examined the earlier contribution by Landsberg (1975) on 'Sahel Drought: Change of Climate or Part of Climate'. In the introduction to this paper, Landsberg asserts that 'it has become fashionable to attribute (the Sahel drought) to climatic change' and in his paper he explores whether such an explanation - that is a change of climate - for the occurrence of the Sahel drought is reasonable. Landsberg then defines his understanding of the term 'climatic change' by stating that 'it implies the shift of climatic conditions overs the globe, and consequently locally, to a *new* equilibrium position'. He further comments that in a climatic change the statistical distribution of values of climatic elements in a climatic change is *significantly changed* in that the data reflect essentially a new population of values. Landsberg concludes that the Sahel drought of the 1970s was not unprecedented and expresses the view that 'it has to be accepted as part of the normal climate of the region'.

The key question of how much 'variation in the climate' would constitute a 'climate change' is further examined by Barring and Hulme (1991); in discussing the Sahelian rainfall from 1900-1988 these authors tentatively suggest that 'a climate change occurs when the 30RU (30 year retrospective running mean with uniform weights) filtered confidence interval does not contain the long term mean value'. They further comment, using the 'definition' of climate change, in relation to the rainfall in the Sahel from 1900 to 1998, that 'through 1985 there had not been any climate change, but if the dry fluctuation continues for only a few more years this would occur'.

Landsberg's viewpoint on climate change is undoubtedly still well respected by many climatologists. Indeed, the more one studies climate, and in particular the climate as indicated by historical records, both instrumental and otherwise, the more one has to ask the questions 'what is climate?', 'what is climate change?', and 'what is 'only' a climate variation?' (For a detailed analysis of climate change from the viewpoint of the Intergovernmental Panel on Climate Change, see Houghton, *et al.,* 1990).

There are of course many 'definitions' of climate, climate change, and climate variations (for a full description of the many organisations, programmes, and terms used in climate, and climate change, see Maunder (1992)), including the definition used by the Intergovernmental Negotiating Committee for a Framework Convention on Climate Change (INC/FCCC), which in its report to the United Nations General Assembly on the 8 May, 1992 provided in its Addendum 1 - for the purpose of the Convention - the following definition of climate change :

'Climate change means a change of climate which is attributed directly or indirectly to human activity that alters the composition of the global atmosphere and which is in addition to natural climate variability observed over comparable time periods'.

I believe, however, that for most if not all climatologists, a more appropriate definition of climate change is that it 'encompasses *all* forms of climatic inconstance (i.e. any differences between the 'long term' statistics of the meteorological elements calculated for different periods but relating to the same area), *regardless* of their statistical nature or physical cause. On the other hand 'human-induced climate change' may be described as a change of climate directly or indirectly induced by human activities and causing significant environmental, economic or social effects. Clearly any convention on climate change needs to address the significance of human 'interference' with the 'natural' regional and global climates. Hence, any Climate Convention definition of 'human-induced climate change' will need to distinguish between natural and anthropogenic, between local and global, and also between significant and insignificant changes of the climate.

Strommen (1993) says that these are major issues that must be resolved and he cites the following key questions for which answers are required:

* Can we distinguish the human-induced signal from that of the natural signal?
* Are the increased greenhouse levels sufficient to alter the natural patterns of climate change and variability?
* Does the observed climate record adequately reflect the expected changes due to the increasing greenhouse gases?
* Is the human contribution to increased levels of greenhouse gases sufficiently great to overcome the causes of the natural variability in the world's climate?
* Are our data good enough to answer these questions?

It is important to note that while the word 'climate' for the purposes of the FCCC is *not* defined; 'climate change' *is* defined, and the definition used by the FCCC of 'climate change' means, in effect, that it is a change of climate which *is due to human activity* that alters the composition of the global atmosphere. Accordingly, it is *not* the same as what could be called a 'natural' climate change. In fact the definition states quite clearly that 'climate change' in the context of the Convention is 'in addition to the natural climate variability which is observed'.

In the Framework Convention on Climate Change it is stated among other things: 'The parties to this Convention acknowledge . . . that change in the earth's climate, and its adverse effects, are a common concern of humankind.' Although this statement includes the phrase 'change in the earth's climate', the word 'change' is not defined, nor is the term 'earth's climate'. One could argue that defining in any legal sense the meaning of either 'change' and/or 'earth's climate', 'climate, climate change', may be unwise. Nevertheless, there has to be a general understanding of what is meant by such terms and for the purpose of this conference it is suggested that a more appropriate definition of climate change - as already suggested - is that it 'encompasses *all* forms of climatic inconstancy (i.e. any differences between the long term statistics of the meteorological elements calculated for different periods but relating to the same area), *regardless* of their statistical nature or physical cause'. Similarly, a climatic variation can be defined as a fluctuation or component thereof whose characteristic time scale is sufficiently long to result in an appreciable inconstancy of successive, medium to short term averages of a meteorological parameter.

While the meaning of the various terms discussed above may be considered at length, I believe that the majority of experts working in the field of climate - whether scientists or advisers - appreciate that climate is indeed 'the fluctuating aggregate of atmospheric conditions characterised by the states and developments of weather of a given area', and that the only real problem is stating what period of time is being used. That is, are we concerned with the climate over the last 10 years, 50 years, 100 years, 1,000 years, or 10,000 years?

From a practical and planning point of view, it is reasonable to suggest that the time interval should be of the order of about 100 years, or possibly a little less. If that is accepted, then a climate change could also be considered to be something that happens *or will happen* on a 'reasonably frequent basis' outside the experience of say the last 100 years.

II. Implications for Nations

At the Second World Climate Conference held in Geneva in November 1989 a number of specific regional and/or national recommendations, proposals and comments were made including the following:

(i) From the Conference Statement from the Scientific/Technical Sessions (The emphasis in *italics* is mine)

* Variations of climate have profound effects on natural and managed ecosystems, the economies of *nations* and the well-being of people everywhere.
* In many cases the impacts (of climate change) will be felt most severely in regions already under stress, mainly in developing *countries.*
* The economic and social costs and benefits of (reducing carbon dioxide emissions) should be urgently examined by all *nations.*
* *Countries* are urged to take immediate actions to control the risks of climate change with initial emphasis on actions that would be economically and socially beneficial for other reasons as well.
* Data acquisition, collection, management and analysis must be more vigorously supported in all *countries* . . . .
* Further development of the methods for predicting short-term variations in climate and the environmental and social impacts should be vigorously pursued (in all *countries*) . . . .
* Projections of future *regional* climate(s) and climate impacts are much less certain than those on a global scale.
* Immediate steps to be taken include . . . *national* and *regional* analyses of the impacts of climate variability and change on society, and study of the range of response and adaptation options available.
* *Governments* . . . should give more emphasis to providing accurate public information on climate issues.

* *Governments* should establish national committees for the World Climate Programme to mobilise support for national activities and to coordinate activities.

(ii) From the Ministerial Declaration

* Recognising climate change as a common concern of mankind, we commit ourselves and intend to take active and constructive steps in a global response, without prejudice to the sovereignty of *states*.
* We maintain that there is a need to intensify research on the social and economic implications of climate change and response strategies. We commit ourselves to promoting the full participation of developing *countries* in these efforts.
* We urge that special attention be given (by *countries*) to the economic dimensions of climate and climate change research.
* The specific difficulties of the those *countries*, particularly developing countries, whose economies are highly dependent on fossil fuel production and exportation, as a consequence of action taken on limiting greenhouse gas emissions, should be taken into account.

It is clear that while many of these recommendations, and proposals/comments are noteworthy, they are understandably not particularly specific, especially from the viewpoint of individual countries which will face many problems both now and in the future, only some of which are climate related. What follows is an attempt to look at the climate and climate change dilemma from the viewpoint of decision-makers in a specific country - which could well be Ireland - the crucial questions concerning competitors both now and in the future, the urgent need to produce national econoclimatic impact studies - considering both the present and any future climates, the controversial issue of winners and losers amongst the 170+ nations of the world, and finally the need to look upon climate change as a challenge and not a threat.

III. Crucial Questions about Competitors

A number of crucial questions need to be asked when discussing the regional implications of climate, climate variations, and climate change. They include the choices regional, national, and local decision-makers have in responding to both climate variations, and a greenhouse gas induced climate change of either their country or countries which are their 'competitors'. In this context, the decisions-makers will be people at local, national, or in the case of some countries in Europe, at the regional level. The real decision-makers could well be someone at the Ministerial level, someone who advises the Minister, or it could be an individual farmer, or manufacturer, or exporter who has to make a climate or climate-change related decision. But, clearly the real decision-maker must be someone who can actually make *and* implement decisions, whether this be the Minister for Finance of Ireland who decides to impose a gasoline fuel carbon tax of 20 pence per litre effective on 1 July 1994, or the owners of a ski company in Switzerland who decide - because of the predicted decreased snow cover in the next 30 years - to sell their ski company and buy instead shares in a solar energy company in Spain.

The term 'competitor' used in the above paragraph could refer to a person, a company or a farm - within Ireland for example - but it can also be defined as a person, a company, or a farm organisation in *another* country with which Ireland is in competition. There are many different areas of competition which may be weather/ climate related, but typical examples include exports and imports, both of which relate directly to the competition in the market place, foreign exchange rates, foreign interest rates relative to local interest rates, and tourism.

In the context of climate variations, and climate change, however, the competitors that need to be considered are *not* only the *present* competitors, but also the ways in which the competitors are likely to change over time. For example, how will international competitiveness in the trade of white wine be affected in the next 30 to 50 years by either the predicted or actual changes in the regional and national atmospheric climate. Moreover, if you believe that the southeastern part of Ireland could become a white wine growing area in direct competition with Germany and New Zealand by the year 2040, because of the anticipated climate change in these areas, what should you do?

In reality, of course, we are not dealing with a simple one crop situation, but rather with a multi-dimensional 'economic-climate chess game' in which the pawns and the kings are likely to change, as well as the 'rules' of climate, economics, and politics. That is, we must deal with the real world of both today and tomorrow.

IV. National Econoclimatic Impact Studies

But first, the following question could well be asked: is the climate of Ireland, and countries which are currently the competitors of Ireland, ideal from Ireland's point of view? From an economic and political viewpoint, there would be some countries - including Ireland, which would argue that a little warming, or a little more rain, would not be a bad thing, and you - as a decision-maker - may be perfectly reasonable in thinking that a little less rain would also not be a bad thing if it occurred to the climate of both Ireland and the countries which are currently your competitors.

Clearly, it would be preferable to be more certain about the magnitude and rate of onset of global, regional, and local weather and climate changes -

particularly as they will affect each of the 170+ countries in the world - and the expected environmental and socio-economic effects. But despite the complexity and scientific and technical excellence of global and regional climate models these uncertainties remain, and are likely to remain for at least some years. However, despite the many uncertainties, there is an urgent need for detailed studies to be made of each of the 170+ countries of the world as to the potential impacts of any future climate changes on their economy and society. This however presents a dilemma, in that except in a few countries, there are few in-depth analyses of the impact of the *past* or the *current* climate on the economy and society of any country in monetary terms (see, for example, Johnson and Holt (1986), Maunder (1970, 1989), Palutikof (1983)).

It is true that several qualitative studies have been made of the impact of the climate (that is the past and/or current climate) such as that made for Australia in 1978 (see CSIRO, 1979), but few if any studies have provided monetary values for the nation as a whole (see Maunder, 1986). Accordingly, useful statements on the monetary value to any nation of its *present* climate are few and far between. It is therefore imperative to assess - as a matter of some urgency - data on the monetary value to each nation, including Ireland, of the value of not only the future climate, but also the present climate, if meaningful decisions are to be made as to how each nation should react to or adjust to any future climate change. Clearly if this is not done, then countries could well be making and will continue to make the wrong decisions.

Since each nation is also in a very competitive real world situation, it is important that economic/climate and economic/climate change budgets be prepared not only for each country, but also - from the viewpoint of each country - for each competitor country.

V. National Winners and Losers

In considering the overall subject of greenhouse gas induced climate change some countries believe that they will be 'winners' and some believe they will be 'losers' with a problem not of their own making. (There are of course 'winners' and 'losers' among the nations of the world when the *present* climate is considered). This belief needs to be taken into account irrespective of political arguments made in some quarters that, overall, the world as a whole will be a 'loser' as a result of the forecast climate change. At present, the information available on climate change impacts is not sufficient to predict winners or losers, but some countries could well perceive that they will be. But, whether countries be 'winners' or 'losers' in a 'warmer world' a number of important issues need to be considered:

* Every nation must be aware of how a climate change will affect it and also how it will affect its competitors, and in order to do this each country

must know the 'value' of the present climate.

* There will be a strong demand for country by country studies, since most communities will want to know how a greenhouse gas climate change will affect their country, and the present and future location of their competitors and competitor industries.

* The role of national political and social groups will need to be recognised. For example, Irish dairy farmers will see a climate change in a completely different way than investors on the Frankfurt stockmarket, or the management of General Motors, and coffee growers in Brazil may well view with concern any trend for competitor coffee growers in other countries to be more favourably affected by any climate change than they are.

* Another key factor in any consideration of climate change is the question: What sort of society can each country - including Ireland, realistically expect in the year 2020, or 2040?, and How should each country react/ adapt to any climate change in the light of the kind of society each country would like to have in the future?

* Irrespective of any climate change (through greenhouse gases or any other cause), climate variations will remain. All countries will continue to be subject to these climate variations (both large and small), and it would be very unwise for any country to place all of their 'climatic eggs' in the 2030, or 2050 basket, and not be equally or even more concerned with what will go into and out of the climate basket during each of the coming years.

* All countries have limited financial and people resources. In considering human-induced climate change, it is important that these resources be directed with appropriate vision to meet the real needs of *each* country. To do this a 'balance sheet' needs to be made of (a) what *has* happened to the climate of each country and the regional and local impacts on each country have been, (b) what *is* actually happening to the climate of each country and the regional and local impacts on each country of what is happening, and (c) what are the current predictions for climate change in each country and the regional and local impacts on each country of these climate changes. Moreover, it is suggested that such 'climate balance sheets' need to be updated on a regular basis, and at least every five years.

VI. The Real Value of Climate to a Nation

Irrespective of the predicted significant climate warming in many areas of the world in the 21st century as a result of the greenhouse effect, many consider that by far the most significant weather and climate variations in the coming years and decades will still occur from week to week, month to month, season to season, and year to year (see Jodha (1990), Phillips (1985), World Meteorological Organisation (1990)). For example, the headline in the Melbourne (Australia) newspaper *The Age* on December 24, 1990 stated: 'It's Christmas in Australia: storms, fires and a cyclone'. The newspaper went on to comment on tropical cyclone Joy, which was threatening northern Queensland, the worst forest fires in New South Wales for 20 years, and severe hailstorms which caused millions of dollars of damage in Melbourne. Greenhouse gases and any potential climate change were in this case a long way from the thoughts of those affected by the three climate-related disasters.

To determine the value that can be placed on these weather and climate-related disasters, together with all other weather and climate variations including floods and droughts, but also the less spectacular but much more common and in many cases much more important economic events such as a mild or cold winter or rainfall 30% above or below average, requires the identification of those economic, social and political activities that are specifically affected by these variations.

In addition, econoclimatic models need to be developed to compute commodity-weighted climate indices for specific areas. Such indices can be assessed by calculating the significance of various sectors - such as retail trade, energy sources, unemployment, or beef production - to the total economic structure of a nation (Maunder, 1986).

Clearly, in any analysis of the impact of climate or climate change on the social, political, and economic structure of any nation, specific econoclimatic studies must be made, and the need for such studies is paramount if decision-makers in each country are to be put in a position where they are able to make climate-related decisions on the basis of facts rather than emotion.

VII. Conclusion

What does the past, present, and future climate mean to each country in the world? Do we have answers to this question. If not, how should we obtain the answers? These are by no means easy questions to answer, but it is imperative that we begin by first assessing in quantitative monetary terms the impact of the past and the present climate on each nation, and then, and only then, attempt to assess in quantitative monetary terms the impact of various projected climate changes on each country for specific time periods such as 2010, 2020, 2030 etc.

As noted in the Introduction, the Second World Climate Conference made several recommendations with regard to national aspects of the climate change problem. Perhaps of most relevance, taking into account the arguments advanced above, are the following:

* Variations of climate have profound effects on natural and managed ecosystems, the economies of *nations* and the well-being of people everywhere.
* Immediate steps to be taken include . . . *national and regional* analyses of the impacts of climate variability and change on society, and study of the range of response and adaptation options available.
* A specific initiative would create a network of *regional* inter-disciplinary research centres, . . . focussing on all . . . disciplines . . . required to support fully integrated studies of global change and its impacts and policy responses.
* *Governments* should establish national committees for the World Climate Programme to mobilise support for national activities and to coordinate activities.

Finally, it is extremely important to note that climate change must be regarded as a challenge, and *not* a threat. Indeed, the challenge is to all scientists, economists, engineers, geographers, entrepreneurs, and politicians, to produce a better environment for all people of all nations. The beginning of that better environment begins with us in the country in which we live.

References

Barring, L., and M. Hulme, 1991. Filters and Approximate Confidence Intervals for Interpreting Rainfall Anomaly Indices, *Journal of Climate,* 4, 963-977.

CSIRO, 1979. *The Impact of Climate on Australian Society and Economy,* Report of a Conference held at Philip Island, Victoria, 27-30 November, 1978, Division of Atmospheric Science, CSIRO, Mordialloc, Victoria, Australia.

Houghton, J.T., G.J. Jenkins, and J.J. Ephramus (Eds.), 1990. Introduction to *Climate Change: The IPCC Scientific Assessment,* Cambridge University Press, xxv-xxxix.

Jodha, N.S., 1989. Potential strategies for adapting to greenhouse warming: perspectives from the developing world, In: Rosenberg, N.J., *et al.* (Eds.), *Greenhouse Warming: Abatement and Adaptation,* Resources for the Future, Washington, D.C. .

Johnson, S.R., and Holt, M.T., 1986. *The Value of Climate Information,* CARD (Centre for Agricultural and Rural Development) Staff Report 86-SR6, Iowa State University.

Landsberg, H.E., 1975. Sahel Drought: Change of Climate or Part of Climate? *Arch. Met. Geoph. Biokl.,* Ser. B, **23,** 193-200.

Maunder, W.J., 1970. *The Value of the Weather*, London, Methuen, pp.388.

Maunder, W.J. 1986. *The Uncertainty Business: Risks and Opportunities in Weather and Climate*, London, Methuen, pp.420.

Maunder, W.J. 1989. *The Human Impact of Climate Uncertainty*, London, Routledge, pp.170.

Maunder, W.J., 1992. *Dictionary of Global Climate Change,* London, UCL Press, pp.240.

Maunder, W.J., 1993. Global Warming: Change of Climate or Part of Climate, *WMO Bulletin*, April 1993.

Palutikof, J., 1983. The Impact of Weather and Climate on Industrial Production in Great Britain, *Journal of Climatology*, **3** , 65-79.

Phillips, D.W., 1985. *Economic Benefits of Climate Applications*, Canadian Climate Program, Atmospheric Environment Service, Downsview, Canada.

Strommen, N.D., 1992. Climate and Crop Yield, *National Geographic Research and Exploration*, **8** (1), 10-21.

World Meteorological Organisation, 1990. Economic and Social Benefits of Meteorological and Hydrological Service, Proceedings of the Technical Conference, Geneva, 26-30 March, 1990, *WMO Publication* No. 733.

CLIMATE AND IRISH HISTORY

John G.Tyrrell
Department of Geography
University College Cork

It may be relevant to ask why the relationship between climate and history should be part of an exploration of policy responses for future climatic variation and change. The answer to this must lie in a connection between the past, present and future in the way climate interacts with society. Despite the particular historical circumstances of specific events of the past, therefore, the task here is to point to generalities in the nature of climatic events, the way in which they impose themselves and the extent to which their impact can be identified and assessed. Where climate and weather have been recognised as having played a significant role in the history of Ireland it has rarely been viewed as a positive one. A greater emphasis has been placed on the hazardous nature of the Irish climate than upon its positive resource value. In part, this has been due to its variability which has made it inconsistent and relatively unpredictable at most time scales.

The many phenomena that make up climate and weather operate over a wide range of time periods. Some are brief and sudden, making an immediate impact upon human society. Other conditions change more slowly and their effects take time to have social and economic consequences. For the purpose of this presentation a distinction is made between three relatively arbitrary time scales because they assist in making important distinctions. The relationship between climate and Irish history can be considered with regard to short,-medium-and long-term climatic effects. The impact of short-term climatic events is usually the most observable and measurable. However, even in the short term the chain of causation can be difficult to trace. With longer time periods the effect of climatic variables is often particularly obscure. But this is not to say that they are, therefore, unimportant and can be discounted. Of these longer time periods, medium-term climatic influences are taken as relating to interannual climatic fluctuations over periods from years to decades, while long-term influences of climate arise from fluctuations and trends which occur on the scale from decades to centuries.

1. Short-Term Impacts

When set in the larger context of social and economic history, the impact of isolated, short-term events has often been small, despite the immediate and often dramatic nature of these events. Sudden brief sharp shocks to the social and economic system, caused by cold and heat, wetness and dryness, together with extreme storms, frosts, floods and droughts have frequently caused damage and claimed lives across the pages of Irish history. Irish society and its economy suffered these impacts but there is little evidence that either changed significantly in response to such individual events. Not that Irish society has been a passive one. Rather, it appears to have been sufficiently resilient to recover, despite the hunger, suffering, disease and mortality in isolated seasons.

At the beginning of the present millenium Irish society was mainly pastoral, though tillage appears to have gradually increased in importance thereafter. The relative unimportance of crops may be part of the reason why few weather events that affected crops were recorded by the early annalists. The most frequently recorded extreme conditions were hard winters associated with frost and snow; their length and severity were important in relation to overwintering cattle herds. In addition, good summers and autumns resulted in welcome additional supplies for overwintering, and were recorded. But the hard winter, the good autumn or other exceptional conditions were isolated, and any food or demographic crisis appears to have been shortlived (Lyons, 1989). Climate, as part of Ireland's resource base, provided a relatively stable condition favourable to both pastoral and cereal production, which was resilient in the face of climatically induced stress. Technologically simple early European societies which were largely self-sufficient are often regarded as having been weather dominated. But this interpretation can be challenged for Medieval Ireland as it has for other parts of Europe (De Vries, 1980). This was probably associated with Ireland's distance from the geographical margin for critical seasonal climatic conditions appropriate for the technology and lifestyle of its people.

Politically and militarily, the period was one in which struggles for the High Kingship gave way to the growth of Norman power, so that by 1250 the Normans controlled nearly 75% of Ireland. Their progress, from the arrival of Strongbow in 1171, was not hindered by extreme conditions as other attempts at invasion were to suffer in later centuries. Immigrant groups of any kind are environmentally dislocated and, as a result, vulnerable in a way that indigenous populations are not. But the Normans do not appear to have encountered difficulties of this nature, either in their military operations or in their subsequent settlement. Indeed, the benign conditions can only have facilitated their superior force, the infrequent occurrence of random hazards providing no opportunity for their opponents to seize the initiative. There have been few turning points in Irish history where the weather can be identified as an overwhelmingly decisive factor in the outcome of events. But the attempted landing in Bantry Bay by the French fleet with Wolfe Tone in

December 1796 was one of these (O'Brien, 1893). Had that landing occurred there was a significant likelihood of Irish history having been different from then onwards. But it was decisively frustrated by a series of severe easterly storms that wrecked and dispersed the French fleet. Short-term individual weather events may have played a role in such key historical situations. However, to dismiss these events (few though they be) merely as chance is to underestimate the significance of the variability characteristic of the climate system (Stolfi, 1980).

2. Medium-Term Impacts

The tendency for extreme conditions of the atmosphere to occur in clusters is a noteworthy feature of many climates. These clusters may have little significance statistically and their physical explanation is often uncertain. However, frequently they have been the most significant expression of climatic variability for human society. They occur as strong departures from normal and persist over a number of years. In particular, pre-industrial societies had coping mechanisms which tended to fail when a cluster of years with extreme conditions put social and economic systems under stress (Post, 1980). The longer the stress was exerted, the greater the likelihood of a crisis that led to disruption and collapse.

Annual clustering of climatic stresses became significant in Ireland from about 1315 onwards. Greater diversity in agriculture had brought with it not just alternative food supplies but also a new vulnerability to climatic conditions that, when they occurred, had been less significant in earlier centuries. So, there was a wider range of hazards at a time when the climate in Europe was 'deteriorating'(Lamb, 1982). From the evidence brought together for Ireland, the fourteenth century was a period of crisis with frequent famine and disease-induced mortality in which persistent extremes of climate are recognised by historians as having played a role (Lyons, 1989). These begin with the Bruce Invasions from Scotland which coincided with a series of very wet winters, summers and autumns between 1315 and 1318, with widespread famine and mortality. Similarly, over the next half century other consecutive years of extreme weather followed which were beyond the population's ability to resist, leading to a major population decline. However, this high frequency of crisis did not persist. The 15th century saw a measure of reversal, although the unhappy coincidence of consecutive weather extremes, poor harvest, warfare and high mortality was to occur from time to time, such as in 1601-1603.

As Irish society and its economy developed into a more complex and modern system from the 17th century onwards, strategies for coping with climatic stresses diversified. This was to change the relationship between society and climate. For example, the potato was introduced as a supplement to cereal-based foods but gained acceptance for succeeding in wet, sunless conditions when the grain crop was poor and when famine would otherwise have threatened (Wood-Martin, 1892). The meteorological differential between potato and cereal crops in wet years was most apparent in Ireland and, as a result, it has been argued that climate played a major role in the establishment of the potato here (Davidson, 1937; Bourke, 1967). The later 19th century crisis was a product of the potato's success, which had led to a total potato diet for many in rural Ireland. Another coping strategy was the development of poor relief. At first this was based on private and local charity but this was followed by co-ordinated administration at city level from 1740 onwards. Emergency food distribution outside normal wholesale channels became important from 1783. In the 19th century government intervention grew in importance, for example with the 1838 Poor Law (Kinealy, 1989).

From the early part of the 19th Century, as the variety of coping mechanisms grew, it is probable that the impacts of climatic stresses became more regionally and socially specific than before. This was by no means an entirely new development. The west of Ireland always had a more pastoral economy than the east and it has been argued that this was something that made it particularly prone to famine when the limited harvests were poor (Gillespie, 1989; Dickson *et al.*, 1981). The early annalists and other sources note that the poor were a particularly vulnerable section of the population, especially their women and children. When succumbing to the effects of famine the cold was frequently mentioned; hypothermia took its toll on vulnerable groups then, as now. By contrast, at the international scale Ireland has been particularly vulnerable in the past during times of continentally wide extremes and associated harvest failures. In such circumstances the possibilities of relieving distress by imports were considerably reduced, as occurred during the 17th and 18th centuries, and the impacts were more generally widespread (Drake, 1968; Post, 1985).

3. Long-Term Change Impacts

The magnitude of long-term changes in climate during the present millenium have been small when expressed as a series of statistical means. Based largely on temperature and to some extent on moisture conditions, Western European climate over this period is usually recognised as having passed through four phases. Evidence to date suggests that the climate of Ireland has broadly reflected these.

These four periods have indicative mean temperature changes which lie within an overall range of 1 - 1.5 degrees celsius. The first of these was the Medieval Period when the climate was warmer than at present. Then, in the early 14th century, decline set in with apparent suddenness (as in Ireland in 1315-18) after a notable period of good summers. This decline was characterised by cooling, wetness and greater storminess. The third period was the Little Ice Age, which was also heralded in by sudden changes in the mid-16th century. The extremes of cold and wetness may have destabilised neighbouring Scotland so much as to lead to large-scale migration in forms ranging from enlisting as mercenaries in the armies of Europe to the establishment of plantations in Ulster (Smout, 1963; Lamb, 1982; Lythe, 1960). Finally, intermittent recovery took place from the 18th century and became strong in the 19th and 20th centuries.

Serious doubts have been raised as to whether such changes influence, in any significant way, a state's agriculture, economy or society. Clearly, the difficulties of isolating the effects of climatic stress are at a maximum at this time scale. The problem is compounded by frequent use of averages, which do not necessarily discriminate between fundamentally different weather and climate conditions, and the tendency for studies to concentrate on single climatic factors, when weather and climate is a complex amalgum of a wide range of atmospheric conditions to which organisms are responding.

Thus, it may be more significant to consider changes in the variability of climate. The climate declines as stressful extremes increase in frequency; it improves as favourable extremes increase in frequency. In contrast to medium-term impacts of climate which affect society through individual clusters of weather conditions, long-term impacts are concerned with the succession of such clusterings over longer periods of time.

The impact of these events appears to have been most evident in marginal areas rather than where weather extremes are expressed in their most intense form. The identity of such marginal areas changes with reference to particular economic enterprises and human activities. Identifying these areas requires not only an assessment of the resource base changes associated with long-term climatic trends, but also the superimposed effects of what Parry has termed proximate social, economic and political factors (Parry, 1978). For example, for the oat crop in Ireland (an important staple over much of the historic past) it is probable that between AD 1300 and AD 1600 there were relatively few truly marginal areas of production and that these were quite small in their geographical extent. The most significant of these was land above approximately 200 metres, including an extensive area around the Wicklow Mountains as well as other areas in the south, west and northern periphery of Ireland where there were strong population concentrations. A more recent impact on the scale of decades, was the changing length of the growing season which resulted in improved conditions for overwintering cattle during the first half of the present century (Lamb, 1982). While the latter was expressed as impacts on the local and regional resource base, viewed in national terms the effect was slight and was not associated with any fundamental change in social or economic conditions.

Conclusion

It may be argued that during different periods over the last 1,000 years climate has been both a destabilising influence and a catalyst for change. Most of the time it appears to have had no significant role that can be identified clearly. But that is to assume that all climatic effects of significance to human society will be direct and apparent. It is clear that climate has been an important part of the total context of both sudden and gradual historical change. Discrimination between the different time-scales of climatic variability and the impact of these on society is a necessary ingredient to identifying the extent to which climate has a forcing role.

The identification and understanding of clusters of significant weather extremes may be a key phenomenon in this task, both for measuring climatic impacts and for developing appropriate policy responses. To do this will require the recognition of the marginality of specific geographical areas and social groups in relation to the range of human activities which underpin our modern society. This suggests that Irish policy responses need to be appropriate both to the particularisms of the Irish resource base, of which climate is an integral part, as well as to the nature of its own society and economy.

References

Bourke, A., 1967. *The Potato, Blight, Weather and the Irish Famine*, Ph.D. thesis, NUI.

Davidson, W.D., 1937. The History of the Potato and its Progress in Ireland, *Journal of the Department of Agriculture,* **34**, 286-307.

Dickson, D., Ó Gráda C., and Daultrey S., 1981. Hearth Tax, Household Size, and Irish Population Change 1672-1821, *Proc.R.Ir.Acad.,* **82** C, 125-181.

Drake, M., 1968. The Irish Demographic Crisis of 1741-41, *Historical Studies,* **6**, 101-124.

Gillespie, R., 1989. Meal and Money: The Harvest Crisis of 1621- 4 and the Irish Economy, In Crawford E.M. (Ed.), *Famine: The Irish Experience,* Edinburgh, John Donald Publishers, 75 - 95.

Kinealy, C., 1989. The Poor Law During the Great Famine: An Administration in Crisis, In Crawford E.M. (Ed.), *Famine: The Irish Experience,* Edinburgh, John Donald Publishers, 157 - 175.

Lamb, H.H., 1982. *Climate, History and the Modern World,* London, Methuen.

Lyons, M.C., 1989. Weather, Famine, Pestilence and Plague in Ireland, 900-1500, In Crawford E.M., (Ed.), *Famine: The Irish Experience,* John Donald Publishers, Edinburgh, 31-74 .

Lythe, S.G.E., 1960. *The Economy of Scotland 1550-1625,* Edinburgh, Oliver and Boyd.

O'Brien, R.B., (Ed.), 1893. *The Autobiography of Theobald Wolfe Tone 1763-1798,* London, T. Fisher Unwin.

Parry, M.L., 1978. *Climatic Change, Agriculture and Settlement,* Studies in Historical Geography Series, Dawson and Hamden, Connecticut, Archon Books.

Post, J.D., 1980. The Impact of Climate on Political, Social and Economic Change: a Comment, *Journal of Interdisciplinary History,* **10**, 719-723.

Post, J.D., 1985. *Food Shortage, Climatic Variability, and Epidemic Disease in Preindustrial Europe: The Mortality Peak in the Early 1740s,* Ithaca.

Smout, T.C., 1963. *Scottish Trade on the Eve of Union, 1660-1707,* Edinburgh, Oliver and Boyd.

Stolfi, R.H.S., 1980. Chance in History: the Russian Winter of 1941-42, *History,* **65**, 214-228.

De Vries, J., 1980. Measuring the Impact of Climate on History: the Search for Appropriate Methodologies, *Journal of Interdisciplinary History,* **10**, 599-630.

Wood-Martin, W.G., 1892. *History of Sligo,* Dublin.

DATASETS FOR CLIMATE CHANGE STUDIES IN THE CLIMATOLOGICAL ARCHIVE OF THE IRISH METEOROLOGICAL SERVICE

Denis Fitzgerald
Meteorological Service

The most comprehensive datasets are collected at the fourteen synoptic stations (see map). Their dates of opening as full synoptic stations, reporting hourly, range from 1939 to the 1960s, with quite a number opening in the 1950s.

Data from all these are available, mostly on an hourly basis, and the elements are:

Temperature (air and soil)
Rainfall
Sunshine
Wind
Humidity
Pressure and pressure tendency
Cloud (height and amount)
Weather (present and past)

Nine participate in a programme of Chemical Analysis of Air and Rainfall.
Six participate in a programme of analysis of Radioactivity Levels.
Seven measure Solar Radiation on a horizontal surface.

All data are quality controlled. The data are held on an Ingres relational database which was acquired in late 1992. Currently we are developing applications and plan to develop new and extended services including user-access to the database, but this is still some time off.

Four of the stations (Birr, Mullingar, Clones and Kilkenny) have operated part-time for the past four years or so; consequently there are gaps in the hourly record for some elements, chiefly cloud, visibility and pressure. This deterioration in the synoptic network is not particular to Ireland but is general and caused by the high cost of manned stations. Use of automatic weather stations, which is on the way here, is seen as the solution to this problem. However, as the sensors are quite different, extreme care is needed during the transition to ensure that the homogeneity of the series is preserved, insofar as this is possible. A period of overlap of about two years, during which readings are taken on both the old and new instrumentation is the recommendation of the international climatological community. More positively, there have been improvements recently in the range of elements read. At Valentia Observatory

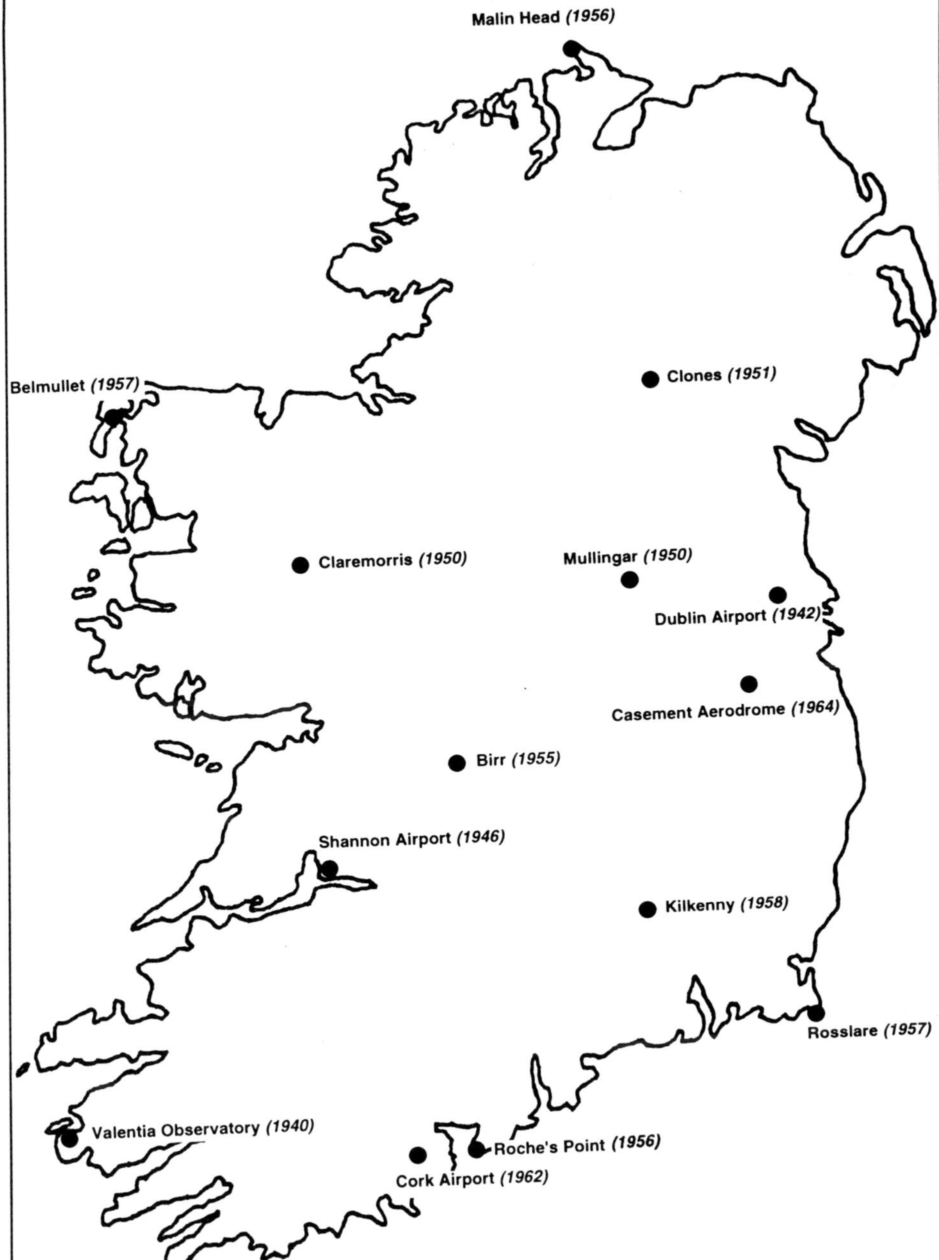

The synoptic observing stations: 1st full year of observations available

soundings of stratospheric ozone are underway and also readings of ultraviolet radiation . . . elements of high interest and concern in recent years.

Rainfall Stations

We receive daily rainfall information from about 650 stations.

Of these, 169 were found in 1988 to have an acceptable reading for each day - a stringent requirement. Indeed there are more than 200 stations which are highly reliable and nearly 600 stations are usable to some degree. Values back to 1941 are available on computer media. Of course there are daily data for the years before 1941, but not on computer. A task for the future is to compile some good series of monthly or even daily data extending back a century of more. In the meantime, a good source of pre-1941 data is the series of publications 'British Rainfall' which are available in our library.

Long-Term Climatological Stations

Climatological stations supplement the synoptic network but their readings are taken daily and not hourly. All read air temperature and rainfall, sometimes sunshine and, more recently, some read soil temperatures and evaporation. At present there are about 70 climatological stations but most were opened since the late fifties. Of course the synoptic stations are climatological stations. However, if by long-term we mean that observations commenced before the turn of the century, the number reduces very sharply. In my view the more important long-term stations are:

Valentia Observatory (Cahirciveen)
Phoenix Park
Malin Head
Markree Castle
Roche's Point
Belmullet-Blacksod
Birr

Valentia Observatory has been at its present site on the mainland since 1892; before that it was located on Valentia Island. Hourly values for various elements have been published for about a century and are available in our library. Unfortunately, values are available on computer only since about 1940; we intend to remedy this situation as soon as possible.

Phoenix Park is the station of longest record, dating back to 1829. However, the Stevenson screen was not introduced here until about 1880 and air temperatures recorded before then must be carefully examined and the exposure of the thermometers determined, if possible. The Campbell-Stokes sunshine recorder was also introduced about 1880 and the exposure of rain-gauges has been in accordance with current practice since that time. The series of values since 1880 constitutes a most valuable series, read in an almost uniform manner - almost, because in the 1940s temperatures were read in a smaller (Bilham) screen instead of the Stevenson screen. The temperature difference between these two types of screen is significant and has to be taken into consideration. This is typical of the traps that await the unwary in climatological work. Provision of metadata, or information about changes of site, instrumentation and exposure, is now a requirement in the international exchange of datasets for climate studies; currently it is a bit more honoured in the breach than in the observance but station history is an integral part of the requirement for producing homogeneous datasets.

Malin Head and Roche's Point have been recording temperature and rainfall since the 1870s. Sunshine has been recorded at Malin Head since 1914 but did not commence at Roche's Point until the mid 50s.

Markree Castle has had continuous records since 1875 but there have been a few hiccups in recent years including the cessation of sunshine recording. Nonetheless it is a valuable record of weather in the north-west; this also applies to the Belmullet-Blacksod record which covers a period of more than a hundred years, but involves a significant change of site. So far very little has been done with these records.

Birr has had observations since 1872 but there have been three significant changes of site, which make it quite a problem to extract a homogeneous series for any weather element.

Monthly values of temperature, rainfall and sunshine are available in manuscript form for Valentia Observatory, Malin Head, Birr and the Phoenix Park for more than a century, but as well as the difficulties mentioned above there is the problem that the hours of reading have varied from 0700 to 2100, rendering it even more tricky to use them as a coherent series.

Developing National Databases

The archive of the Irish Meteorological Service is a valuable national resource. However, its value to the nation could be enhanced by judicious combination with datasets held by other agencies, e.g. a national hydrological database could be formed from parts of the Meteorological Service's archive (rainfall, evaporation), EPA-ERU (river flow, water quality) and the Geological Survey (groundwater). The advantage for the researcher is obviously that one request would suffice where previously three might have been needed. For the co-operating agencies the advantage is that the number of requests for data is reduced provided, of course, that they are not in charge of the integrated database - a separate agency could be set up for this but it would have to be funded in some way. There is also the surrender of a possible source of income in times when all are enjoined to take a commercial view. Most of these difficulties can be met by sharing any income from the database or simply reverting to a broader perspective where considerations of the national good prevail over narrower considerations.

Other Important Datasets

Upper Air

Vertical soundings of the atmosphere are taken only at Valentia Observatory and measurements of winds, temperature and humidity dating back to the late 1940s are available.

Synoptic Maps

The British Meteorological Office started publishing synoptic surface weather charts in the Daily Weather Report in 1872. Our library in Glasnevin has numbers dating back to the early 30s while Valentia Observatory has copies back to 1915. However, sets of surface weather charts dating back over 100 years are now available on microfilm.

The BMO started publishing an upper air section from 1919, but the main body of upper air data dates from the 1940s and is available in manuscript form in our library. In the 1980s the BMO ceased publication of the Daily Weather Report, but we are now archiving the Deutscher Wetterdienst publication which has a lot of surface and upper air information.

Marine

We have data from lighthouses, oil rigs, ships and buoys, most of it fairly short-term from the point of view of climate studies. An exception is the ship information which dates back about 100 years and is a most important dataset even if great care must be exercised in using it.

CLIMATE VARIABILITY IN IRELAND: ARE THERE SIGNALS AMONGST THE NOISE?

Stuart Daultrey
Department of Geography
University College Dublin

Introduction

Although climate has always been recognised as incorporating variability, most emphasis has usually been placed on climatic means and normals. The global climatic classification systems of Köppen and Thornthwaite developed from the late nineteenth century through to the first half of this century rely entirely on mean monthly temperature and precipitation data (Hare, 1951). More attention has been paid to variabilty in the second half of this century with the development of a dynamic approach to climatology, exemplified in Ireland and Britain by Hubert Lamb's daily airflow classification (Lamb, 1972) and the development of synoptic climatology (Barry and Perry, 1973). This dynamic approach has revealed that much climatic variability is not random, but a function of large and often complex ocean-atmosphere interactions that produce marked interannual oscillations. The best known of these is the El Niño/Southern Oscillation (ENSO), which affects the weather and climate not just of the South Pacific but of almost the entire globe (Yarnall, 1985; Yarnall and Kiladis, 1985). Other major causes of variability in the Northern Hemisphere are the North Atlantic Oscillation (Lamb and Peppler, 1987), the North Pacific Oscillation (Rogers, 1981) and the Quasi-Biennial Oscillation (QBO) of equatorial stratospheric winds on its own (Parker, 1982) and in conjunction with the approximately 11-year sunspot cycle (Labitzke and van Loon, 1988).

If accurate prediction is to extend beyond the paltry five days currently on offer from the meteorologists there must be continued and intensified observation and modelling of these ocean-atmosphere (and -cryosphere) interactions and their effects. And as if that weren't complex enough, there must be added the effects of human activity altering the composition of the atmosphere and the nature of the surface, both affecting the overall energy balance and both with the potential to alter not just the mean state of the climate system but also its variability (Parker and Folland, 1988). A preliminary analysis of the relationships between the variability of one aspect of the Irish climate, winter precipitation, and three atmospheric oscillations appears elsewhere (Daultrey, 1994). This paper focuses on the issue of identifying secular changes in Irish climate that might be caused by human modification of atmospheric composition.

Modification of Atmospheric Composition by Human Activity

Since the middle 18th century the concentration of carbon dioxide (CO_2) in the atmosphere has risen from approximately 270 ppmv to its current 360 ppmv. This has been caused by deforestation, predominant until the late 1950s, and the burning of fossil fuels, now the predominant source. Carbon dioxide is a so-called 'greenhouse gas', meaning that it is relatively transparent to incoming short-wave radiation from the sun but highly absorbent of outgoing long-wave radiation from the Earth and atmosphere. Increasing the concentration of any 'greenhouse gas' increases the absorption of heat in the lower atmosphere, where most of it will be re-radiated to the Earth's surface. The result is that more heat energy is concentrated at the Earth's surface and in the lower atmosphere at the expense of the upper atmosphere. All other things being equal the surface and lower atmosphere temperatures should rise, and the upper atmosphere temperatures should fall: every additional $W.m^{-2}$ concentrated in the lower atmosphere is approximately equivalent to a 1 K rise in temperature (Kiehl and Dickinson, 1987). There are other 'greenhouse gases', principally methane (CH_4), nitrous oxide (N_2O) and many chlorofluorocarbons (CFCs) whose atmospheric concentrations have been increased by human activity. While their concentrations are orders of magnitude less than that of carbon dioxide, their capacities for absorption of long-wave radiation are orders of magnitude higher, so that their combined effect is about 80% of that of carbon dioxide. The effective CO_2 concentration is therefore in the range 390-420 ppmv (Shine *et al.*, 1990), an increase in lower atmosphere radiative forcing of 2-2.5 $W.m^{-2}$ (Hansen and Lacis, 1990). So, all other things being equal, global temperatures should have risen by 2-2.5 K since 1750.

There are currently no reliable estimates of global temperature prior to 1850, but there is a variety of data sets containing global and hemispheric annual estimates since that date. Using autoregressive (AR) and autoregressive integrated moving average (ARIMA) models, Bloomfield (1992) calculates the rate of increase for the 1861-1989 data set prepared for the IPCC (Folland *et al.*, 1990) as 0.37-0.38°C per century, slightly less than the IPCC's own estimate of 0.45°C (IPCC = Intergovernmental Panel on Climate Change). Bloomfield's calculations for the 1881-1987 data assembled by Hansen and Lebedeff (1988) show rises of 0.55-0.58°C per century, the difference being due mainly to the urban warming bias in the Hansen and Lebedeff data and to their non-inclusion of the years 1861-1880, when there was little or no warming. Woodward and Gray (1993), while not disputing these findings, point out that the correlation structure of both data sets precludes the acceptance of any hypothesis other than that these rates of increase are short-term. In other words, the observed data alone do not confirm the existence of global warming. Balling and Idso (1990), Idso (1990) and Michaels and Stooksbury

(1992) point to the fact that most of the warming took place prior to 1945, when the increase in the concentration of 'greenhouse gases' was one third of the total increase. Since 1945, with two thirds of the 'greenhouse gas' enhancement, the Northen Hemisphere showed a slight reduction in temperatures until 1975, and then a rapid increase, while the Southern Hemisphere showed a continued increase in temperature although perhaps at a reduced rate. Even if the observed data do indeed indicate a global warming trend, there need be no reason to argue any further than a climatic recovery from what is called the Little Ice Age, a cooler period from the fifteenth to the nineteenth centuries that followed the Medieval Warm Epoch, a warmer period from the tenth to the thirteenth centuries (Idso, 1989). The average temperature differences between these two periods in Europe of at least 1.5°C are larger than any changes being observed at present. However, recent work has suggested that the global extent and uniform coolness of the Little Ice Age are doubtful (Bradley and Jones, 1992). Nevertheless, Kozuchowski (1993) shows significant positive correlations, strong in winter, between a zonal index of the strength of the North Atlantic westerlies and European temperatures over the period 1899-1990, suggesting that 'greenhouse' warming (which should result in weakened Atlantic westerlies) is unimportant to climate variability over Europe.

As Hulme (1993) points out, detection of an enhanced 'greenhouse effect' requires not just the identification of significant climatic change but also the attribution of that change to an enhanced greenhouse effect as opposed to any other forcing mechanism. That attribution can only be made by climate models, and these are described in the IPCC reports (Cubasch and Cess, 1990; Gates *et al.*, 1990, 1992). The models of the early to middle 1980s tended to give estimates of global warming of around 4 K in response to a doubling of atmospheric CO_2 from 300 to 600 ppmv, with the high latitude winters showing most warming and the tropics least. Later models, incorporating deep oceans, improved cloud parameterisations, finer meshes and transient responses, tend to produce warming estimates of about 2 K, still with warming concentrated in high latitude winters, but with reduced warming or even cooling over the northern North Atlantic and in the oceans around the Antarctic. While such results more closely resemble the actual pattern of temperature change than do those of earlier models, they still do not anticipate, for example , the observed cooling (1967-86) over most of Europe and northern South America (Jones, 1988), and their estimates of warming are still at least double the observed increases. However, as Duffy (1993) points out, this does not mean that model predictions of major greenhouse warming must be wrong.

Climate models do not incorporate the effects of other human-induced changes to the atmosphere, and they do not incorporate the natural variability such as might arise from volcanic emissions or solar influences, whereas the observed temperature record must include these effects. Izaksen *et al.* (1992) focus on the effects of two human-induced changes, the emission of CFCs and of sulphate aerosols. CFCs are highly effective 'greenhouse gases', but they have greater notoriety for their ability to destroy ozone (O_3). Ozone is also a 'greenhouse gas', but it is far more important to the global energy balance in its role as an absorber of very short-wave radiation (ultra-violet wavelengths <0.29 μm) in the stratosphere. Ozone depletion because of CFCs causes stratospheric cooling which in turn alters the surface-troposphere radiation balance to promote cooling in the lower atmosphere. Ramaswamy *et al.* (1992) found that net CFC forcing may be negative in middle and high latitudes and positive in low latitudes, reducing their global 'greenhouse' forcing by 80% and hence effective CO_2 forcing by 12%. Sulphate aerosols released by the burning of fossil fuels, besides creating the well-known acid rain problems, act both as direct reflectors of short-wave radiation and as cloud condensation nuclei which particularly promote low cloud formation over oceans. In each case the Earth's albedo is increased, giving a net cooling. Hansen and Lacis (1990) estimate aerosol forcing of -1.5 $W.m^{-2}$, but Charlson *et al.* (1992) give an upper limit of -2 $W.m^{-2}$ for sulphate forcing equally divided between the direct albedo effect and the indirect cloud albedo effect. As most of the sulphate emissions occur in the Northern Hemisphere, these estimates are more than enough to nullify the 2-2.5 $W.m^{-2}$ 'greenhouse gas' forcing mentioned earlier and therefore explain the lack of temperature rise since 1945, an argument also pursued by Idso (1990). However, the rapid increase in sulphate emissions occurred from the 1950s, by which time Northern Hemisphere cooling was well-established, and the argument would appear to fail to account for the marked Northern Hemisphere warming since 1975. Kaufman and Chou (1993) model the temperature response to CO_2 forcing and to CO_2+SO_2 forcing according to the IPCC's Business as Usual (BAU) scenario and their D scenario (a much lower rate of fossil fuel consumption; see Houghton *et al.*, 1990). Their findings indicate that the indirect cloud albedo effect is considerably greater than the direct albedo effect, that SO_2 forcing reduces the BAU scenario 1990-2060 warming from 1.9 to 1.4°C and the D scenario warming from 1.1 to 0.9°C, and that the (negative) SO_2 forcing in the Northern Hemisphere remains 0.2°C greater than that in the Southern Hemisphere over the period 1985-2060. Interestingly, an upward inflection in temperatures in the model runs occurs at 1980 with a magnitude corresponding closely to observational evidence.

Asymmetric Temperature Trends

In the 1980s the work of Thomas Karl and his associates revealed a general trend in North American temperatures of minimum temperatures increasing much faster than maxima, that this decrease in daily temperature range was linked to changes in precipitation and especially to increases in cloudiness, and that these changes appeared to be related to increasing concentrations of 'greenhouse gases' (Karl *et al.*, 1984, 1986; Plantico *et al.*, 1990). The analysis was extended to China, the former Soviet Union, Japan and parts of northern Europe, and data were also collected from the Sudan, South Africa and eastern Australia, giving coverage of 50% of the Northern Hemisphere land surface and 10% of the Southern Hemisphere (Karl *et al.*, 1993), summarised in columns B-D, table 1.

Michaels and Stooksbury (1992, p.1574) articulated the proposed relationships in the following manner. If low-level cloudiness due to human activity were increasing in a climatologically significant fashion, the following should be observed. Firstly, night warming should appear because of increases in both 'greenhouse gases' and cloudiness. Secondly there should be a counteraction of daytime warming because of increased cloud albedo, thereby decreasing daily temperature range. Thirdly, the greatest warming effect of clouds should occur on long winter nights. Fourthly, the greatest counteraction of warming should occur on long summer days. Fifthly and sixthly, the least warming effect and least counteraction of warming should occur respectively on short summer nights and short winter days. Seventhly, cloudiness should be enhanced near the North American and Eurasian source regions of cloud condensation nuclei. Finally, these effects should be concentrated in the Northern Hemisphere. They then aggregated the Karl *et al.* (1991) data for the USA, China and the former Soviet Union, reproduced here as column A of table 1, to confirm the first five expectations, and also produced evidence to support the last two. Inspection of the hemispheric and global averages of Karl *et al.* (1993), reproduced in table 1, reveals that although in very general terms the expected relationships hold, there are important differences. For example, the magnitudes of the winter and spring warmings are much greater in the latter data set. Also, spring night warming is greater in both hemispheres than winter night warming, and the counteraction of day warming is in both hemispheres slightly greater in the autumn than in the summer. There is obviously at least one other factor operating besides daylength and cloudiness, because as defined in these studies spring daylengths are longer than autumn daylengths.

Temperature and Sunshine Trends in Ireland

The data used in the analyses of Karl *et al.* (1993) were collected for the period 1951-1990. Of the Republic of Ireland's synoptic weather stations, only five have uninterrupted and consistent data for that period, namely: Claremorris, Co. Mayo; Clones, Co. Monaghan; Dublin Airport, Co. Dublin; Shannon

Airport, Co. Clare; and Valentia Observatory, Co. Kerry. To these five were added the Northern Ireland station at Belfast International Airport, Aldergrove, Co. Antrim. Mean monthly maximum and minimum temperatures, monthly mean temperatures and mean daily temperature ranges, and monthly averages of daily hours of bright sunshine, were aggregated into means for the calendar year and for four three-monthly seasons (winter - Dec-Feb; spring - Mar-May; summer - Jun-Aug; autumn - Sep-Nov). The data relating to mean annual temperatures, mean annual daily temperature ranges and mean annual daily hours of bright sunshine for the six stations are plotted in figures 1-3 as annual deviations from the 1961-1990 means together with nine-year weighted running means 'padded' at each end of the time period.

Recent temperature trends (°C per century)

	A		B		C		D	
Period	**Max**	**Min**	**Max**	**Min**	**Max**	**Min**	**Max**	**Min**
Winter	0.6	1.8	1.3	2.9	1.6	2.2	1.3	2.9
Spring	0.6	1.5	2.0	3.2	1.7	2.5	1.9	3.1
Summer	-0.4	0.4	-0.3	0.8	1.0	1.3	-0.2	0.8
Autumn	-0.6	0.7	-0.4	1.3	0.8	2.1	-0.3	1.4
Year	0.05	1.1	0.5	2.0	1.3	2.0	0.7	2.1

Table 1. Recent temperature trends. **A)** Area weighted combination of USA, former Soviet Union and People's Republic of China (Michaels and Stooksbury, 1992). **B)** 50% of Northern Hemisphere land area. **C)** 10% of Southern Hemisphere land area. **D)** 37% of global land area. (**B-D** 1951-1990, from Karl *et al.*, 1993).

The values for the six stations were averaged to give seasonal and annual estimates for Ireland as a whole. The change in values over time was described by simple linear regression coefficients, expressed in °C per century for the temperature data and hours per century for the sunshine data. Coefficients significant at the 10% level or less are shown in bold; those significant at the 5% level or less are also underlined. Table 2 shows the coefficients for Ireland as a whole, table 3 for the three western stations (Claremorris, Shannon Airport, Valentia Observatory) and table 4 for the three eastern stations (Belfast Aldergrove, Clones, Dublin Airport).

Recent temperature trends in Ireland (six station mean)

Period	Ave	Max	Min	Range	Sun
Winter	2.0	1.3	**2.8**	**-1.5**	**-1.1**
Spring	-0.8	-1.1	-0.5	-0.6	**-1.5**
Summer	**1.7**	1.8	**1.6**	0.2	-1.3
Autumn	0.3	0.1	0.5	-0.5	-0.3
Year	0.8	0.5	**1.1**	-0.6	**-1.0**

Table 2. Regression coefficients showing rates of change for Ireland, 1951-1990, of average, maximum, minimum temperatures, temperature range and sunshine hours. For further details, see text.

Recent temperature trends in western Ireland stations

Claremorris

Period	Ave	Max	Min	Range	Sun
Winter	1.9	1.5	2.4	**-0.9**	**-2.0**
Spring	-0.6	-0.5	-0.8	0.3	**-3.1**
Summer	**1.6**	2.1	1.2	0.8	**-3.1**
Autumn	-0.4	-0.2	-0.6	0.4	**-1.0**
Year	0.5	0.6	0.4	0.1	**-2.3**

Shannon Airport

Period	Ave	Max	Min	Range	Sun
Winter	**3.5**	1.7	**5.3**	**-3.6**	**-1.4**
Spring	0.6	-1.0	**2.2**	**-3.2**	**-1.8**
Summer	**1.3**	1.4	**3.2**	**-1.8**	**-2.5**
Autumn	0.9	-2.0	2.0	**-2.2**	-0.6
Year	**2.7**	0.3	**3.0**	**-2.7**	**-1.6**

Valentia Observatory

Period	Ave	Max	Min	Range	Sun
Winter	1.6	1.3	1.8	-0.5	**-1.8**
Spring	-1.4	-1.3	-1.5	0.2	**-2.4**
Summer	0.4	0.6	0.1	0.5	**-2.5**
Autumn	-0.8	-0.4	-1.2	0.7	-0.6
Year	-0.2	-0.1	-0.3	0.2	**-1.8**

Table 3. Regression coefficients showing rates of change for three stations in western Ireland, 1951-1990, of average, maximum, minimum temperatures, temperature range and sunshine hours. For further details, see text.

For Ireland as a whole, there has been a statistically significant rise in annual minimum temperatures at a rate of 1.1°C per century compared to a statistically non-significant rise in annual maximum temperatures of 0.5°C, giving a (statistically non-significant) fall in the annual daily temperature range. There has also been a statistically significant fall in annual sunshine at a rate of 1.0 hrs per century. If the fall in sunshine can be taken as a surrogate for a rise in cloudiness, then the first two expectations of Michaels and Stooksbury (1992) are confirmed. The rise in minimum temperatures is greatest in winter, in line with their third expectation, but the counteraction of the rise in maximum temperatures is least in the summer in direct contradiction to their fourth expectation. Expectation five, of least night warming on short summer nights, is not supported, and neither is expectation six of least day cooling on short winter days. The general pattern revealed by table 2 is of statistically significant falls in sunshine in winter and spring with the expected faster increase of minima over maxima in winter, but an unexpected faster decrease of maxima over minima in spring. Statistically non-significant falls in sunshine in summer and autumn are accompanied by increases in maxima and minima in summer, but autumn appears to be a season of little change.

Recent temperature trends in eastern Ireland stations

Belfast Aldergrove

Period	Ave	Max	Min	Range	Sun
Winter	**2.8**	2.1	**3.5**	**-1.3**	0.3
Spring	-0.0	-0.6	0.5	-1.1	-0.5
Summer	**1.8**	1.7	**1.9**	-0.2	0.3
Autumn	-0.1	-0.4	0.3	-0.7	0.9
Year	**1.0**	0.6	**1.4**	**-0.8**	0.3

Clones

Period	Ave	Max	Min	Range	Sun
Winter	2.0	1.2	**2.7**	**-1.4**	**-1.1**
Spring	-0.8	-1.1	-0.5	-0.6	-1.0
Summer	**2.1**	2.3	**1.9**	0.4	0.4
Autumn	0.2	-0.1	0.5	-0.6	0.3
Year	0.8	0.6	**1.1**	-0.5	-0.3

Dublin Airport

Period	Ave	Max	Min	Range	Sun
Winter	**3.2**	**2.5**	**3.8**	**-1.3**	**-0.8**
Spring	0.5	0.8	0.3	0.5	-0.9
Summer	**2.0**	**2.2**	**1.7**	0.6	-1.2
Autumn	1.0	1.1	0.9	0.2	0.0
Year	**1.5**	**1.5**	**1.5**	0.0	**-0.7**

Table 4. Regression coefficients showing rates of change for three stations in eastern Ireland, 1951-1990, of average, maximum, minimum temperatures, temperature range and sunshine hours. For further details, see text.

The six stations were selected only because of the completeness of their data. There is no reason to suppose that the averages of the six are completely representative of climatic variability and trends over Ireland. Indeed, three of them are airport sites and thus would probably not have been selected for any representative station network such as those of Karl *et al.* (1993). When the individual station trends are described there become apparent major differences between the three stations located in western Ireland (table 3) and three located in the east (table 4). The three western stations show consistently large and statistically significant decreases in hours of bright sunshine in all seasons except autumn, whereas the three eastern stations show no statistically significant change other than small decreases in winter sunshine at Clones and Dublin Airport. However, while the three eastern stations show statistically significant changes in temperature in winter, summer and in the annual means, two of the western stations, Valentia Observatory and Claremorris, show none. This leaves Shannon Airport as the only one of the six stations to exhibit both statistically significant reductions in sunshine (and, by implication, increases in cloudiness) and statistically significant changes in temperature.

Conclusions

The pattern of temperature changes, with minima increasing faster than maxima in winter, maxima increasing faster than minima in summer, and weak and variable changes in spring and autumn, is reasonably consistent across all six stations. The problem is to explain this pattern in terms of observed changes in other climatic parameters, in terms of observed changes in patterns of ocean and atmosphere circulation, and in terms of the changes in energy balance brought about by the effects of human activity on the composition of the Earth's atmosphere and on the nature of the surface.

There is some consistency between the observed changes and enhanced cloudiness due to sulphate aerosols. But it is quite possible that sulphate emissions will be drastically reduced in the near future in order to overcome the problems of acid rain. The model of Kaufman and Chou (1993) suggests that were this to happen there would be very rapid global warming, 0.4°C in less than a decade. However Michaels (1993) suggests that while sulphate aerosols may be assisting in the enhanced cloudiness and the asymmetric temperature trends, they are ultimately a product of 'greenhouse warming' itself. He points to the tropospheric temperature records of Angell (1991) which show that temperatures in the 300-100 mb layer have been declining. He speculates that if the necessary upper atmospheric cooling to compensate for lower atmospheric warming takes place from 300 mb upwards, and not from 100 mb upwards as is the conventional wisdom, then the increased tropospheric lapse rates would promote cloud formation and produce a more energetic but not necessarily much warmer lower atmosphere. He concludes that this 'probably spells the end of the greenhouse scare' (Michaels, 1993, p.857).

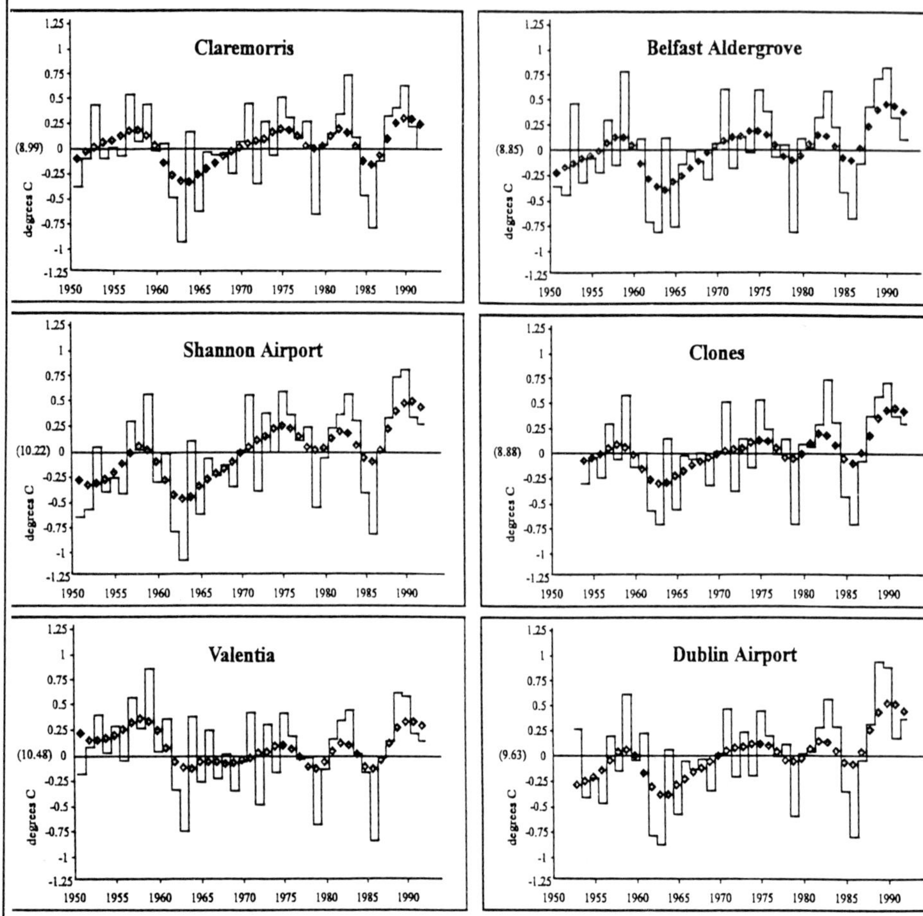

Figure 1. Mean annual temperatures, 1951-1992, for six stations in Ireland. Annual values are shown as deviations from the 1961-90 mean by the vertical bars. The shaded diamonds are weighted 9-year running means 'padded' to fill both ends of the series.

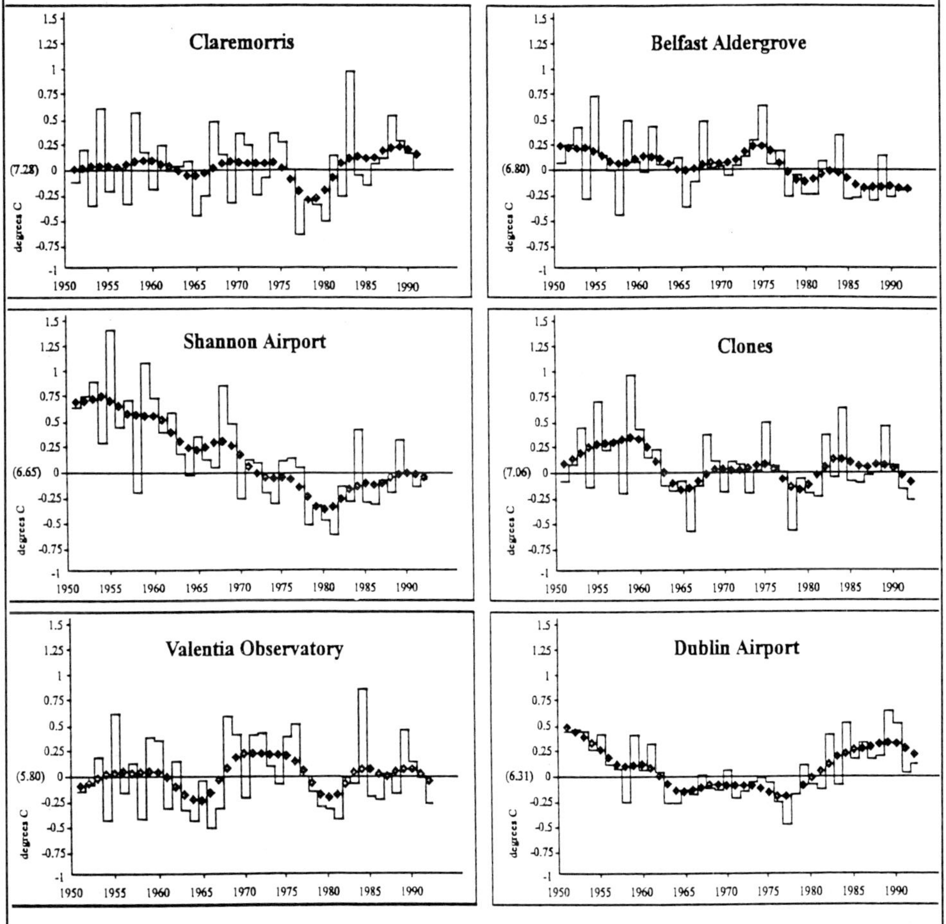

Figure 2. Mean annual daily temperature ranges, 1951-1992, for six stations in Ireland. Annual values are shown as deviations from the 1961-90 mean by the vertical bars. The shaded diamonds are weighted 9-year running means 'padded' to fill both ends of the series.

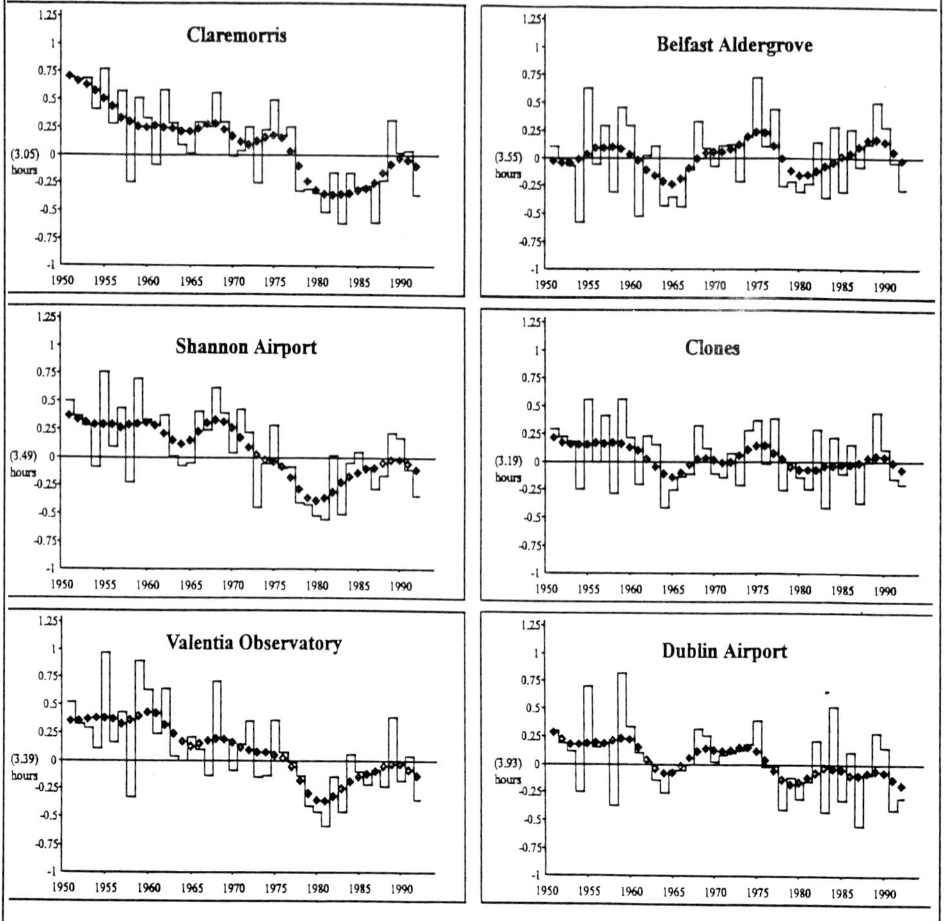

Figure 3. Mean annual daily hours of bright sunshine, 1951-1992, for six stations in Ireland. Annual values are shown as deviations from the 1961-90 mean by the vertical bars. The shaded diamonds are weighted 9-year running means 'padded' to fill both ends of the series.

References

Angell, J.K., 1991. Changes in tropospheric and stratospheric global temperatures, 1958-88, In Schlesinger, M.E., (Ed)., *Greenhouse-gas-induced climate change: A critical appraisal of simulations and observations*, Amsterdam, Elsevier, 231-247.

Balling, R.C. and Idso, S.B., 1990. 100 years of global warming? *Environmental Conservation*, **17**, 165.

Barry, R.G. and Perry, A.H., 1973. *Synoptic climatology: Methods and applications*, London, Methuen.

Bloomfield, P., 1992. Trends in global temperature, *Climatic Change*, **21**, 1-16.

Bradley, R.S. and Jones, P.D. (Ed.), 1992. *Climate since AD 1500*, London, Routledge.

Charlson, R.J., Schwartz, S.E., Hales, J.M., Cess, R.D., Coakley, J.A., Jr., Hansen, J.E. and Hoffmann, P.J., 1992. Climate forcing by anthropogenic aerosols, *Science*, **255**, 423-430.

Cubasch, U. and Cess, R.D., 1990. Processes and modelling, In Houghton, J.T. *et al.* (Eds.), *Climate change: The IPCC scientific assessment*, Cambridge, Cambridge University Press, 69-91.

Daultrey, S., 1994. The influences of the North Atlantic Oscillation, the Quasi-Biennial Oscillation and the El Niño/Southern Oscillation on winter precipitation in Ireland, In Jones, J.A.A. and Woo, M.-K., Eds., *Proceedings of IGU Washington and Lhasa Symposia on regional response to climate change*, forthcoming.

Duffy, P.B., 1993. Comments on "Global warming: A reduced threat?" by P.J. Michaels and D.E. Stooksbury, *Bulletin of the American Meteorological Society*, **74**, 855.

Folland, C.K., Karl, T.R. and Vinnikov, K.Ya., 1990. Observed climate variations and change, In Houghton, J.T. *et al.* (Eds.), *Climate change: The IPCC scientific assessment*, Cambridge, Cambridge University Press, 195-238.

Gates, W.L., Mitchell, J.F.B., Boer, G.J., Cubasch, U. and Meleshko, V.P., 1992. Climate modelling, climate prediction and model validation, In Houghton *et al.* (Eds.), *Climate Change 1992: The supplementary report to the IPCC scientific assessment*, Cambridge, Cambridge University Press, 97-134.

Gates, W.L., Rowntree, P.R. and Zeng, Q.-C., 1990. Validation of climate models, In Houghton, J.T. *et al.* (Eds.), *Climate change: The IPCC scientific assessment*, Cambridge, Cambridge University Press, 92-130.

Hansen, J.E. and Lacis, A.A., 1990. Sun and dust versus greenhouse gases: An assessment of their relative roles in global climate change, *Nature*, **346**, 713-718.

Hansen, J. and Lebedeff, S., 1988. Global surface air temperatures: Update through 1987, *Geophysical Research Letters*, **15**, 323-326.

Hare, F.K., 1951. Climatic classification, In Stamp, L.D. and Wooldridge, S.W. (Eds.), *London essays in geography*, London, London School of Economics and Political Science, 111-134.

Houghton, J.T., Callander, B.A. and Varney, S.K., (Eds.), 1992. *Climate Change 1992: The supplementary report to the IPCC scientific assessment*, Cambridge, Cambridge University Press.

Houghton, J.T., Jenkins, G.J. and Ephraums, J.J. (Eds.), 1990. *Climate change: The IPCC scientific assessment*, Cambridge, Cambridge University Press.

Hulme, M., 1993. Global warming, *Progress in Physical Geography*, **17**, 81-91.

Idso, S.B., 1989. Greenhouse warming or Little Ice Age demise: A critical problem for climatology, *Theoretical and Applied Climatology*, **39**, 54-56.

Idso, S.B., 1990. Evidence in support of Gaian climate control: Hemispheric temperature trends of the past century, *Theoretical and Applied Climatology*, **42**, 135-137.

Izaksen, I.S.A., Ramaswamy, V., Rodhe, H. and Wigley, T.M.L., 1992. Radiative forcing of climate, In Houghton *et al.* (Eds.), *Climate Change 1992: The supplementary report to the IPCC scientific assessment*, Cambridge, Cambridge University Press, 47-67.

Jones, P.D., 1988. Hemispheric surface air temperature variations: Recent trends and an update to 1987, *Journal of Climate*, **1**, 654-660.

Karl, T.R., Jones, P.D., Knight, R.W., Kukla, G., Plummer, N., Razuvayev, V., Gallo, K.P., Lindseay, J., Charlson, R.J. and Peterson, T.C., 1993. A new perspective on recent global warming: Asymmetric trends of daily maximum and minimum temperature, *Bulletin of the American Meteorological Society*, **74**, 1007-1023.

Karl, T.R., Kukla, G. and Gavin, J., 1984. Decreasing diurnal temperature range in the United States and Canada from 1941-1980, *Journal of Climate and Applied Meteorology*, **23**, 1489-1504.

Karl, T.R., Kukla, G. and Gavin, J., 1986. Relationship between decreased temperature range and precipitation trends in the United States and Canada, 1941-1980, *Journal of Climate and Applied Meteorology*, **25**, 1878-1886.

Karl, T.R., Kukla, G., Razuvayev, V.N., Changery, M.J., Quayle, R.G., Heim, R.R., Jr., Easterling, D.R. and Fu, C.B., 1991. Global warming: Evidence for asymmetric diurnal temperature change, *Geophysical Research Letters*, **18**, 2253-2256.

Kaufman, Y.J. and Chou, M.-D., 1993. Model simulations of the competing climatic effects of SO_2 and CO_2, *Journal of Climate*, **6**, 1241-1252.

Kiehl, J.T. and Dickinson, R.E., 1987. A study of the radiative effects of enhanced atmospheric CO_2 and CH_4 on early earth surface temperatures, *Journal of Geophysical Research*, **92**, 2991-2998.

Kozuchowski, K.M., 1993. Variations of zonal hemispheric index since 1899 and its relationship with air temperature, *International Journal of Climatology*, **13**, 853-864.

Labitzke, K. and van Loon, H., 1988. Associations between the 11-year solar cycle, the QBO, and the atmosphere. Part I: The troposphere and stratosphere in the Northern Hemisphere in winter, *Journal of Atmospheric and Terrestrial Physics*, **50**, 197-206.

Lamb, H.H., 1972. British Isles weather types and a register of the daily sequence of circulation patterns, 1861-1971, *U.K. Meteorological Office Geophysical Memoir*, **116**.

Lamb, P.J. and Peppler, R., 1987. The North Atlantic Oscillation: Concept and an application, *Bulletin of the American Meteorological Society*, **68**, 1218-1225.

Michaels, P.J., 1993. Reply [to Duffy]. *Bulletin of the American Meteorological Society*, **74**, 856-857.

Michaels, P.J. and Stooksbury, D.E., 1992. Global warming: A reduced threat? *Bulletin of the American Meteorological Society*, **73**, 1563-1577.

Parker, D.E., 1982. Large-scale interannual variability of climate, *Meteorological Magazine*, **111**, 193-208.

Parker, D.E. and Folland, C.K., 1988. The nature of climatic variability, *Meteorological Magazine*, **117**, 201-210.

Plantico, M.S., Karl, T.R., Kukla, G. and Gavin, J., 1990. Is recent climatic change across the United States related to rising levels of anthropogenic greenhouse gases? *Journal of Geophysical Research*, **95**, 16617-16637.

Ramaswamy, V., Schwartzkopf, M.D. and Shine, K.P., 1992. Radiative forcing of climate from halocarbon-induced global stratospheric ozone loss, *Nature*, **355**, 810-812.

Rogers, J.C., 1981. The North Pacific Oscillation, *Journal of Climatology*, **1**, 39-57.

Shine, K.P., Derwent, R.G., Wuebbles, D.J. and Morcrette, J.-J., 1990. Radiation forcing of climate, In Houghton, J.T. *et al.* (Eds.), *Climate change: The IPCC scientific assessment*, Cambridge, Cambridge University Press, 41-68.

Woodward, W.A. and Gray, H.L., 1993. Global warming and the problem of testing for trend in time series data, *Journal of Climate*, **6**, 953-962.

Yarnal, B., 1985. Extratropical teleconnections associated with El Niño/Southern Oscillation (ENSO) events, *Progress in Physical Geography*, **9**, 315-352.

Yarnal, B. and Kiladis, G., 1985. Tropical teleconnections associated with El Niño/Southern Oscillation (ENSO) events, *Progress in Physical Geography*, **9**, 524-558.

BASELINE CLIMATE OBSERVATIONS AT MACE HEAD ATMOSPHERIC RESEARCH STATION

Aodhagán F. Roddy *, S. Gerard Jennings & Thomas C. O'Connor
Department of Physics, University College Galway
*also, Meteorology Section, Martin Ryan Marine Science Institute, University College Galway

1. Introduction

Some atmospheric measurements were made at Mace Head as early as 1958. The earliest measurements were on the concentrations of Aitken nuclei and other aerosol particles, and were made at a small rented lookout post not far from the site of the current Atmospheric Research Station. Similar measurements, together with some other research, were carried out sporadically during the 1960s and the 1970s. An opportunity was grasped in 1975 to purchase an 8 acre coastal site with a small cottage in ruins. The cottage was refurbished in 1985 for use as a laboratory, and some standard meteorological measurements were undertaken on a continuous basis.

In 1987 the cottage laboratory became one of the sites for the Global Atmospheric Gases Experiment (GAGE). This enabled the continuation and extension of the pioneering work of Dr James Lovelock on the measurement of CFC gases, which he had carried out at Adrigole, another remote site in the south-west of Ireland. With the extension of the atmospheric research at Mace Head and with wider participation, another milestone was reached in 1989 when a new laboratory was constructed close to the shoreline. This laboratory was built in co-operation with the Atmosphere/Ocean Chemistry Experiment (AEROCE) project. A 20 m sampling tower was also erected in conjunction with the AEROCE project. Later in 1989, the Government provided funding, under the National Science and Technology Development Programme, for the construction of a second shoreline laboratory, and an extension to the (original) cottage laboratory. The Minister for Science and Technology, Mr Michael Smith, performed the opening ceremony in the summer of 1990 when the construction of these additional facilities was completed.

2. The Site and Facilities at Mace Head

The Atmospheric Research Station at Mace Head is situated on the west coast of Ireland, approximately 80 km west of Galway City, near the village of Cárna, Co. Galway (see Fig. 1). There are a few small islands to the west of the station which are not inhabited. The site was originally chosen because it is in a remote area on the Atlantic coast of Europe, in the path of the frequent mid-

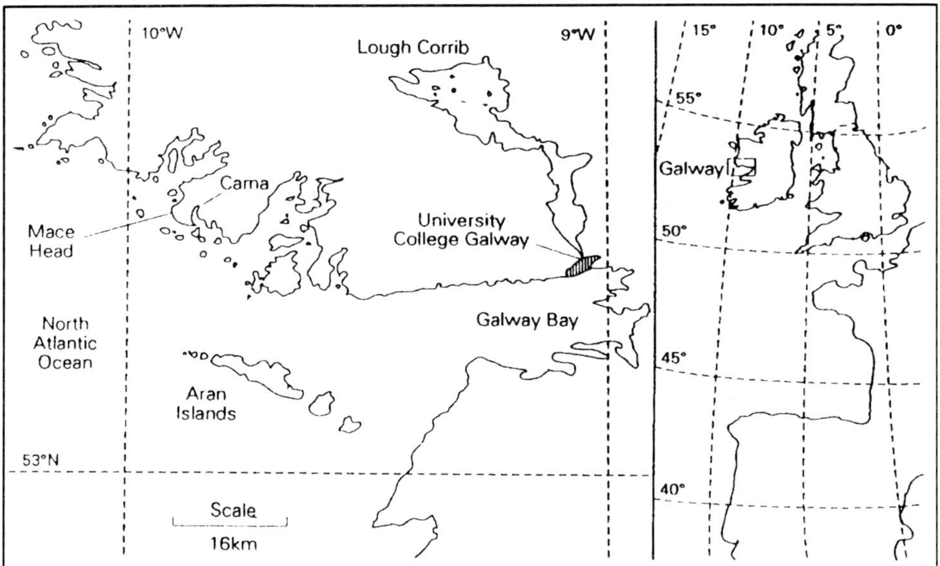

Fig. 1: Location of the Atmospheric Research Station at Mace Head

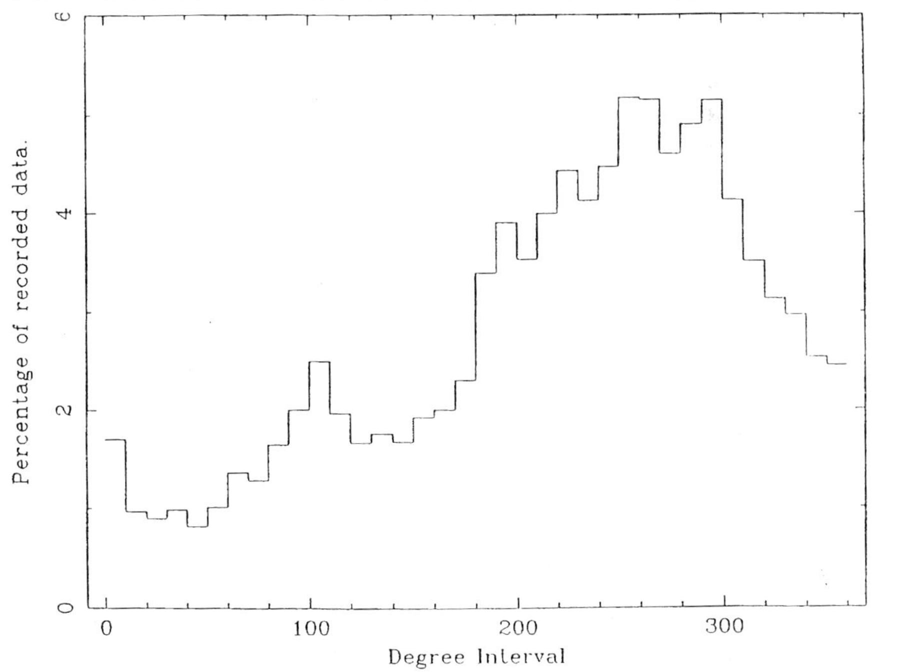

Fig. 2 : Wind direction data at Mace Head for 1990

latitude weather systems which, for over 60% of the time, bring in air with a recent trajectory over the ocean and which is generally not contaminated by local anthropogenic pollution (see Fig. 2). The station has a nominal 120° clean sector between the local directions of 180° and 300°. Fig. 3 shows carbon (in aerosol) concentration as a function of wind direction for the month of November, 1990. This shows clearly the main sector sources of anthropogenic influence at the site for that month. Thus, the site is very suitable for research on such topics as global background trace gases and aerosol, natural biogeochemical cycles and air-sea exchange processes. But occasional significant pollution events also occur at the site, when, typically, European continental air, originating from an easterly direction, reaches Mace Head. The study of such events yields valuable information on topics such as the long-distance transport of pollutants from highly industrialised areas.

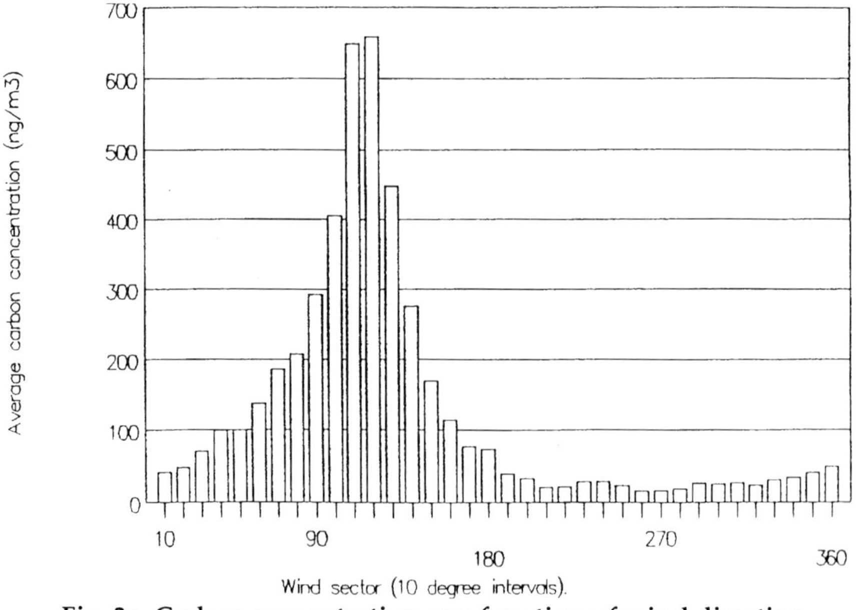

Fig. 3 : Carbon concentration as a function of wind direction, November, 1990

At present there are three laboratory buildings, namely the two shoreline laboratories and the cottage laboratory, on the Mace Head site. Facilities include the 20 m sampling tower, a 10 m meteorological tower and platforms for the positioning of sampling equipment.

3. Programmes and Projects

Research programmes and projects at Mace Head include the following:

* The US Atmosphere/Ocean Chemistry Experiment (AEROCE) on biogeochemical cycles in the North Atlantic, which is part of the North Atlantic Regional Experiment in the International Geosphere Biosphere Programme (IGBP). Among those involved in this programme are groups from several US Universities and the National Oceanic and Atmospheric Administration (NOAA).

* The Global Atmospheric Gases Experiment (GAGE) which involves studies on the global distribution of the principal 'greenhouse gases'. These include methane, surface ozone, nitrous oxide, and the main chloroflourocarbon (CFC) gases. Research groups from the USA, Australia, and the UK are participating in this programme.

* An investigation of trace pollutant levels in the Irish atmosphere, involving also the setting up of a database for a range of micropollutants. This project is carried out in collaboration with St Patrick's College, Maynooth, and is funded, through the Department of the Environment, by the STRIDE programme of the European Union .

* An investigation of the background maritime contribution to atmospheric pollution in Europe. The project involves the study of specific chemical and microphysical properties of the predominately maritime air masses in their passage from the Atlantic Ocean over Western Europe. This work is being carried out in collaboration with institutes in Portugal, France, UK, Sweden and Norway and is funded under the EU Environment Programme.

* The measurement of concentrations of surface ozone and some precursor gases such as carbon monoxide and oxides of nitrogen (NO_x), as part of the EUREKA / EUROTRAC Tropospheric Ozone Research project (TOR).

* The measurement of atmospheric carbon dioxide, in collaboration with CNRS, France. This EU funded programme also involves C_{13} /C_{12} isotope measurements.

* Ultraviolet radiation measurements, in co-operation with the Irish Meteorological Service. This work forms part of a European network programme and is funded through the Department of the Environment.

* Flask measurements of carbon dioxide, carbon monoxide, and methane, as part of the NOAA / CMDL Co-operative Flask Sampling Network.

* Studies of mercury and other heavy metal concentrations in the air, carried out in collaboration with GKSS and the University of Hanover.

* The investigation of natural sources of sulphur compounds such as dimethyl sulphide (DMS) and sulphur dioxide, involving collaboration with the Max Planck Institute at Mainz.

4. Species Measured at Mace Head

Species measured at Mace Head include the following:

Trace Gases:
CO_2, CH_4, N_2O, O_3, CO, CFC 11, CFC 12, CFC 113, CCl_4, CH_3Cl, NO_x, NO_y, NMHCs, SO_2, DMS, Rn^{222}.

Particulate Measurements:
Aitken nuclei, cloud condensation nuclei (CCN), aerosol size and mass distributions, aerosol volatility. Total suspended particulates, size-segregated particulate samples. High-volume samples for analysis of major ions, trace elements, sulphur isotopes and Pb^{210}/Be^7 ratios. Particulate carbon.

Precipitation Measurements:
Major ions, trace elements, total wet deposition.

Meteorological Measurements:
Wind direction and speed, temperature, relative humidity, rainfall, pressure, solar radiation.

5. Some Examples

Extracts from two sets of measurements are illustrated below, in order to give an impression of two of the programmes underway at Mace Head.

CFCs and some other gases:
Fig. 4 shows trends in measurements of CFC 11, CFC 12, nitrous oxide, methyl chloroform and carbon tetrachloride over a period of years. Measurements of some of these species had been carried out by James Lovelock at a remote site at Adrigole in the south-west of Ireland, for a number of years in the late 1970s and the early 1980s, but these measurements were discontinued for some time. They were commenced again at Mace Head, together with additional measurements, in January, 1987, as part of the GAGE programme, (See for example Prinn et al., 1992; O'Connor & Simmonds, 1991).

Methane:
Atmospheric methane has been measured at Mace Head, as part of the GAGE programme, since 1987. The methane is measured by means of a Hewlett Packard 5880 gas chromatograph, at approximately two-hourly intervals. Fig. 5 shows these basic measurements of methane concentration, as a function of time, for the month of June, 1990. There appears to be a background level of about 1780 ppbv for much of that month. Fig. 6 shows

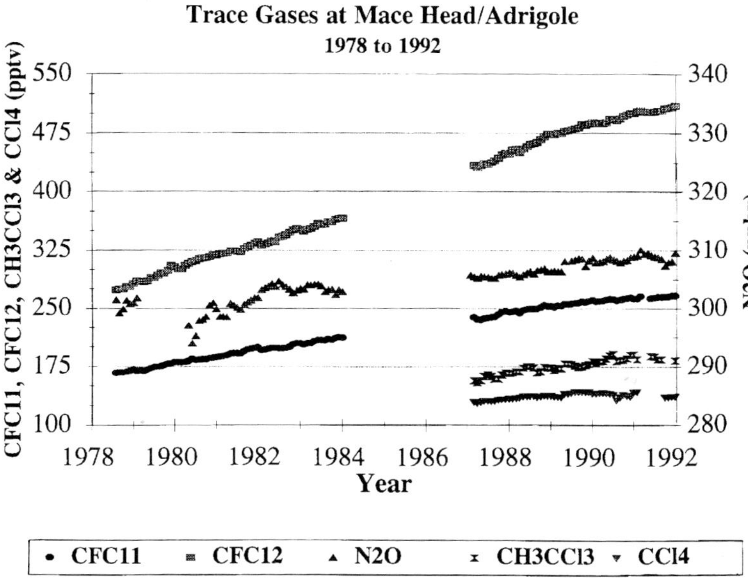

Fig. 4 : Trends shown in measurements of some trace gases at Mace Head

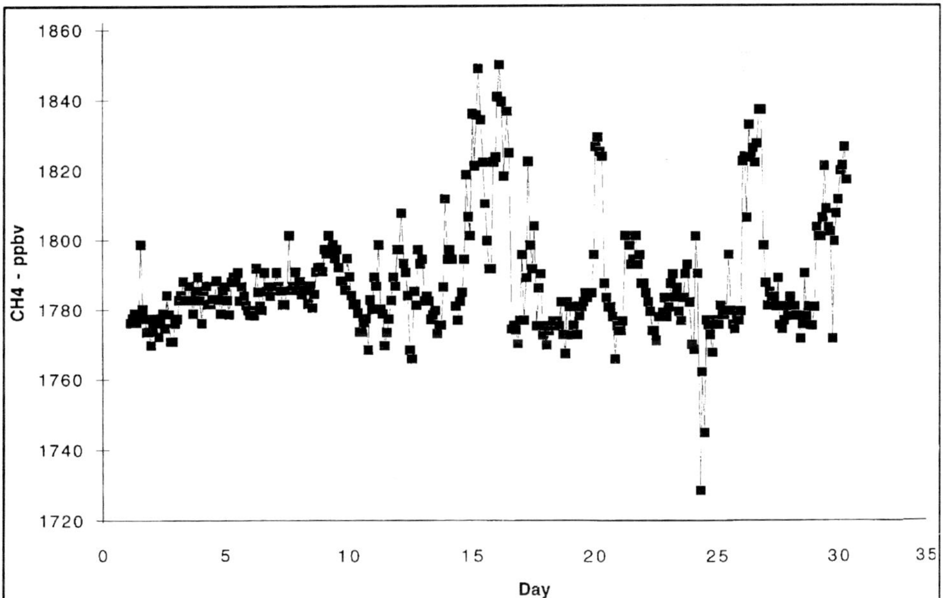

Fig. 5 : Atmospheric methane measurements at Mace Head, June, 1990

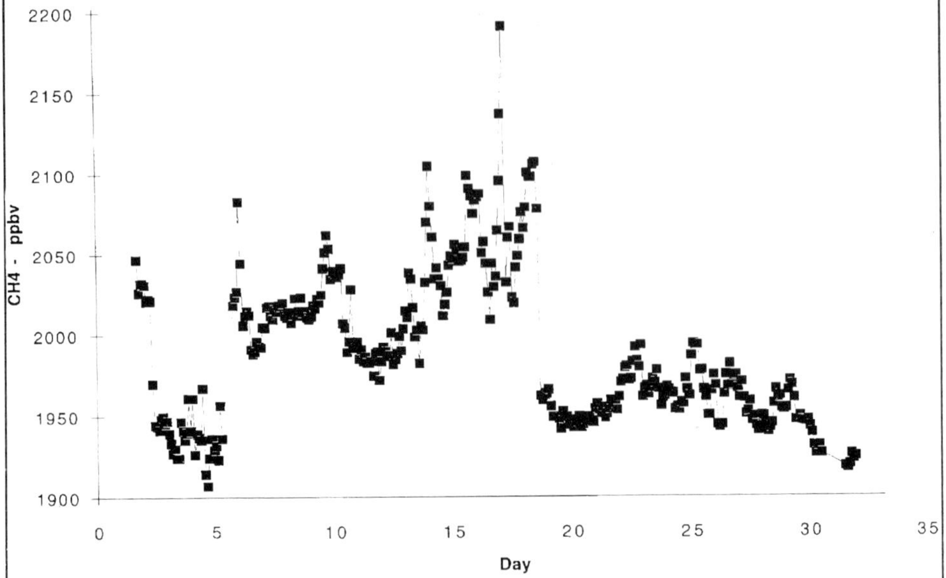

Fig. 6 : Atmospheric methane measurements at Mace Head, December, 1990

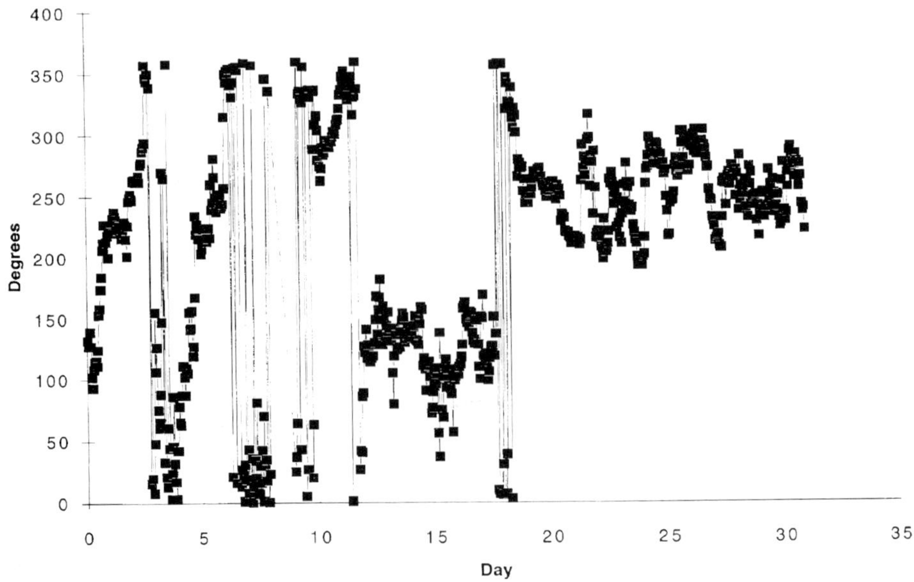

Fig. 7 : Wind direction measurements at Mace Head, December, 1990

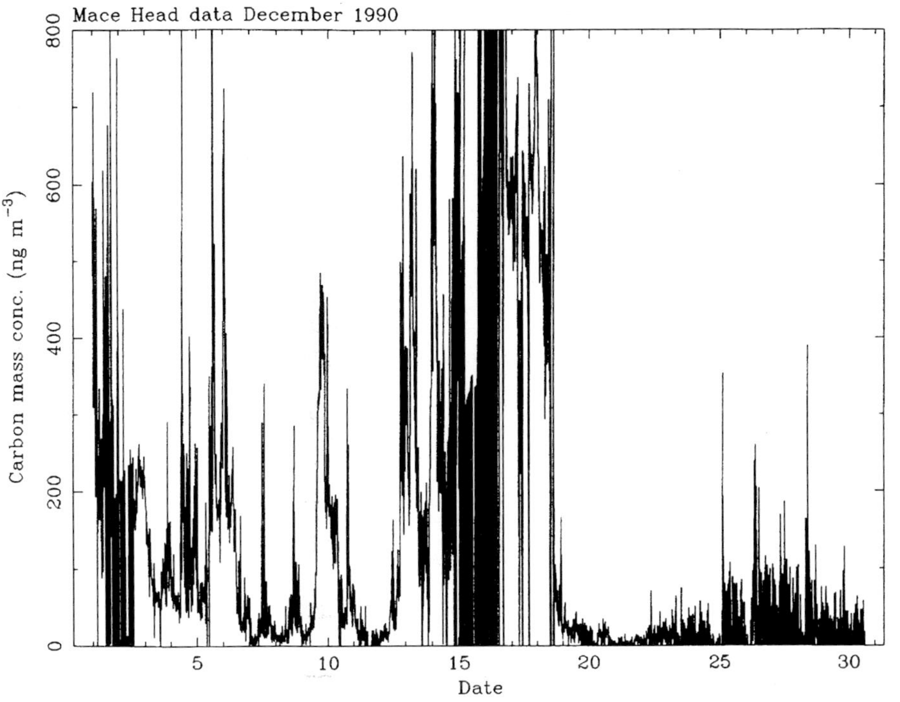

Fig. 8 : Carbon mass concentrations at Mace Head, December, 1990

methane concentrations for the month of December, 1990. These December values are obviously at a higher level; at no stage during December did the concentration of methane fall below 1900 ppbv. For a period of days in mid-December, from about the 14th to the 19th December, methane concentrations reached as high as 2100 ppbv on several occasions, and nearly 2200 ppbv on one occasion. Fig. 7 shows a plot of wind direction v. time for December, 1990. For the days of particular interest, 14th - 19th December, there was a significant easterly component, with a change to a northerly direction towards the end of that period. Fig. 8 shows a plot of carbon mass concentration versus time for December 1990. Here, again, there is evidence for a polluted air mass, particularly for the period of days around the middle of the month, from about the 14th to the 19th December. The wind direction is not neccessarily a good guide to the origin and/or history of the air mass. Air mass trajectory analysis, however, can give a good indication with regard to the more recent history of the air. Therefore, isentropic air mass back trajectory analyses were obtained for air arriving at Mace Head for periods of interest (Merrill, 1993). Fig. 9 shows the back trajectory analysis for 00 UTC on the 16th December. This indicates that the air passed over northern

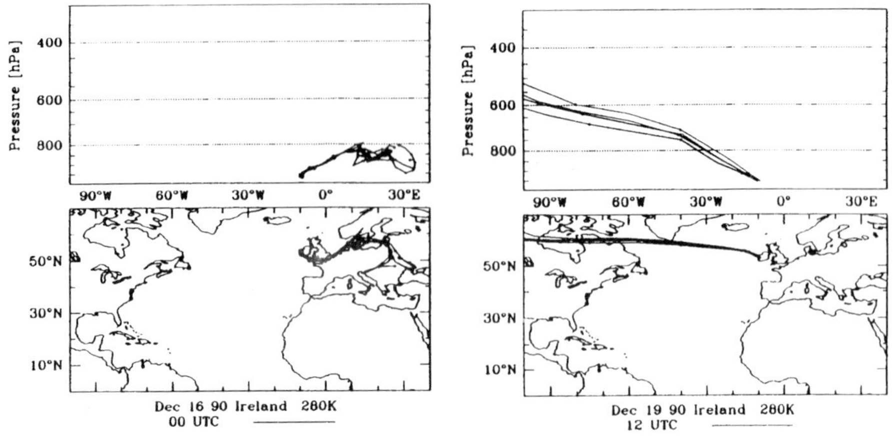

Fig. 9 : Isentropic back trajectories for air arriving at Mace Head at 00 UTC, 16th December, 1990 and 12 UTC, 19th December, 1990

Europe and the south of England, prior to its arrival at Mace Head. Similar analyses are available for other days in the period 15th to 19th December, 1990. However, on the 19th December, the situation changed completely, and the air arriving at Mace Head came from a westerly direction, and had passed over the Atlantic Ocean for some time, as shown by the back trajectory for 12 UTC in Fig. 9. Thus, on particular occasions, even a very remote part of Ireland can come under the influence of a highly polluted air mass.

Acknowledgements

Support from the Irish Government, the STRIDE and Environment programmes of the European Union, AEROCE, GAGE and several other programmes is acknowledged. Other members of the Atmospheric Physics Research Group have assisted in various ways.

References

Merrill, J. T. (University of Rhode Island) 1993. Private communication.

O'Connor, T.C. and Simmonds, P., 1991. CFCs in Conamara, *Technology Ireland*, **23**, 19-21.

Prinn, R., Cunnold, D., Simmonds, P., Alyea, F., Boldi, R., Crawford, A., Fraser, P., Gutzler, D., Hartley, D., Rosen, R. and Rasmussen, R., 1992. Global average concentration and trend for hydroxyl radicals deduced from ALE/GAGE trichloroethane (methyl chloroform) data for 1978-1990, *J. Geophys. Res.*, **97**, 2445-2461.

CLIMATE MODELLING

Peter Lynch
Meteorological Service
Dublin

Present climate is determined from analysis of meteorological observations. Climates of the past may be deduced by studying proxy data such as ice-cores, tree-rings, pollen, etc. Future climate changes may be estimated by means of computer simulations using programs known as General Circulation Models or Climate Models.

People have built physical models having some properties in common with the atmosphere. These 'rotating dish-pans' do contain wave motions and instabilities similar in some respects to those observed. But no-one has constructed a mechanical model which behaves in detail like the atmosphere. The difficulties with such a task are obvious: for example, we cannot replicate the radial gravitational force of the spherical earth in the laboratory. So, we design instead a mathematical model in which the geometry of the atmosphere can be precisely represented and a variety of detailed physical processes incorporated. The computer program to solve or integrate this system of mathematical equations is our climate model.

The essence of modelling is to use the known laws of physics to design a computer program capable of simulating the atmospheric flow. With realistic initial observational data the program can produce a short-range weather forecast, typically for up to a week ahead. If the solution is carried forward for an extended period - months or years of simulated flow - the forecast will diverge rapidly from the true evolution of the atmosphere. However, its statistical characteristics - mean values, frequency of extremes, etc. - may resemble those of the real climate provided the model includes a representation of all the processes which determine climate.

The model has as its basis the fundamental principles of physics - conservation of mass and energy and Newton's laws of motion. These determine the overall behaviour of the atmosphere. Many physical processes must also be allowed for: the phase changes of water, incoming solar radiation, frictional drag at the earth's surface, sub-grid-scale turbulence and so on. The details of some of these microphysical processes are poorly understood with the consequence that there are inherent inaccuracies and uncertainties in all climate models.

In studying climate change, we must also include chemistry in the model. Small changes in the concentration of some species may dramatically alter the radiation balance and profoundly influence the climate. Details of the interaction of solar and terrestrial radiation with compounds such as methane and ozone must be taken into account. A vast complex of chemical reactions occurs in the atmosphere. Current models have relatively simple treatment of chemical processes. Future models may need to consider the concentrations of hundreds of compounds, their reaction rates, and so on.

The most advanced climate models are capable of simulating the observed climate with a good level of accuracy. The zonal mean winds and temperatures in each season are generally well simulated, and the statistics of extremes are realistic. Thus, a reasonable picture of the climate of a given geographical location emerges from the models. However, it must be acknowledged that there are still serious shortcomings. In a recent comparative study of 14 climate models, it was found that the temperature was in all cases too cold, particularly in the polar upper troposphere and tropical lower troposphere. It was conjectured that all the models are misrepresenting or even omitting some mechanism, resulting in this deficiency.

So, we can simulate the current climate fairly faithfully. But why bother? Why spend large resources finding out what we already know? Well, our fond hope is that a climate model can do more than simulate the status quo. Like a laboratory model, the climate model can be used in experiments to study how the climate may change in the future. For example, we may study the sensitivity of model climate to a doubling of carbon dioxide simply by changing a single number (representing CO_2 concentration) and re-running the model. This may give us an indication of what is likely to happen in the real world under such circumstances. Indeed, just such experiments are the basis of current predictions of global warming resulting from the burning of fossil fuels. However, model results have been found to vary widely when details of the physical parameterisations are adjusted: in recent experiments with the Hadley Centre model, the predicted increase in global mean surface temperature ranged from 1.9°C to 5.2°C using different cloud schemes. The following quotation is from a report on these experiments:

> 'All present-day GCM results must be treated with caution. The models contain many approximations, and hence uncertainties which, through a lack of understanding of the underlying physical and chemical processes of the climate system, cannot (yet) be quantified and/or removed'.

The guidance provided by model sensitivity studies is about the best we have for now. But there is some danger in placing too much reliance on it. The equations governing the atmosphere are non-linear, their solutions are hyper-sensitive to small perturbations and minor changes can have major implications. Thus, seemingly negligible deficiencies in a climate model may render its output useless or misleading. Moreover, there are certainly physical and chemical processes which are insignificant in present conditions and are therefore ignored in current models, but which may be of critical importance in altered circumstances. As an example, the detailed micro-physics of ice clouds is involved in the depletion of Antarctic ozone. This was not foreseen by the modellers, with the result that we were all caught on the hop. I fear there may be more unpleasant surprises to come.

In conclusion, the overall success of climate models in simulating the present climate of the atmosphere is impressive. Although there are shortcomings in all models, they give a generally accurate picture of reality. They provide a valuable means for estimating the likely climatic consequences of changes induced by mankind's activities. At present, we must interpret their guidance with some caution; in particular, the details of geographical variations in climate impact may be unreliable. We can be confident that the dependability of model guidance will grow with their increasing sophistication, so that detailed regional climate impact predictions may be reliable in the future, perhaps shortly after the turn of the millennium. But in an uncertain world we can never completely rule out those 'nasty surprises'.

References

(1) Washington, W.M., and Parkinson, C.L., 1986. *Basic Text: An Introduction to Three-dimensional Climate Modelling,* University Science Books.

(2) *Recent CO2 Sensitivity Study.* The Hadley Centre Transient Climate Change Experiment, Hadley Centre Publication, August, 1992.

(3) Boer, G. J. *et al.,* 1991. *Recent Model Intercomparison Study: An Intercomparison of Climates Simulated by 14 Atmospheric General Circulation Models,* CAS/JSC WGNE, WMO T.D. No.425, July.

IMPLICATIONS FOR IRELAND OF CLIMATE CHANGE - AN OVERVIEW

Brendan E. McWilliams
Assistant Director
Meteorological Service

General

The *Environment Action Programme* published in January 1990 by the Minister for the Environment proposed the commissioning of a series of Expert Studies to establish the impact of possible climate change on specific aspects of the Irish environment. The studies were carried out by invited experts and published by the Department of the Environment under the title *Climate Change - Studies on the Implications for Ireland* in late 1991.

Six topics were chosen:

- The impact of climate change on agriculture
- The impact on forestry
- The Irish green mantle and its native fauna
- Hydrology and freshwater resources
- The impact on Ireland of changes in mean sea level
- Fisheries and shell fisheries in Irish coastal waters

The Studies did not address the likelihood or otherwise of climate change itself. It was *assumed* that the Irish climate *would* change, and the objective was to try to quantify the impact of this hypothetical change in an Irish context. They focused on the period between 1990 and the year 2030, by which time it was also assumed that an equivalent CO_2 doubling would have taken place.

The contributors were provided with a number of working assumptions, broadly in line with the consensus reached by Working Group I of the Intergovernmental Panel on Climate Change (IPCC), which had recently been published in *Climate Change - The IPCC Scientific Assessment* (July 1990). Specifically, the authors were asked to assume that the following changes would take place in Ireland before the year 2030:

Average Annual Temperature	**- an increase of the order of 2°C**
Precipitation	**- 5 to 10% increase in winter**
	- 5 to 10% decrease in summer
Mean Sea Level	**- an increase of 18 cm**

They were also asked to assess possible 'best' and 'worst' scenarios, being given appropriate assumptions in each case.

It was realised that more sophisticated General Circulation Models might in due course render the assumptions invalid, but for the purposes of these Studies they represented a best guess on which to assess hypothetical impacts and plan possible response strategies. Moreover, all the experts emphasised the tentative nature of their conclusions, and the high level of uncertainty inherent in speculations of the kind required for the Studies.

Agriculture

The study on Agriculture was prepared by Teagasc at the Johnstown Castle Research Centre in Co. Wexford. It was concluded that in general the agricultural production potential of Northern Europe would be enhanced if the climate were to change along the lines envisaged. Ireland would share in this advantage - perhaps even to a disproportionate extent if prolonged summer droughts did not become a problem. The production options available would be increased, new crops could be cultivated, and the overall costs of agricultural production would be likely to be less than is the case at present.

Grass yields were likely to increase by up to 20% because of the higher temperature and changed rainfall patterns, and perhaps by another 10% due to increased CO_2 in the atmosphere. This enhanced growth would have significant implications in many related areas of agricultural activity - e.g. the length of grazing season, grass conservation, over-wintering, and the housing of farm livestock. Livestock farmers would be able to grow other forage crops like maize, fodder beet, lucerne and red clover, which would lead to the development of new grassland management systems, as well as to a significant reduction in nitrogen fertiliser use.

Little or no increase in cereal yields was foreseen. Yields of other tillage crops like sugar beet and potatoes, however, might increase by up to 20%, although pests and weeds might pose a greater problem than at present.

Arable crops currently grown predominantly in the south and midlands might find northern regions equally amenable in the new scenario; similarly the cultivation of a number of new crops - such as maize, sunflower, and flax - might become viable over much of the south and south-east.

Significant changes would be likely in horticulture. The cultivation of French beans, haricot beans, courgettes, summer cauliflower, celery and early potatoes could become more common, while apples, pears and cherries could be grown over wider areas than is currently the case - assisted by a reduction in the frequency of late spring frosts. The successful harvesting of gooseberries and raspberries west of the Shannon would become possible.

The principal agricultural input industries, on the other hand, might expect a reduced demand for their products, despite the predicted increases in agricultural output. The feed and fertiliser industries would probably be the most adversely affected. Processors of livestock products, however, should benefit from the increased throughput and stronger demand for their goods.

The changes envisaged were likely give Ireland a competitive advantage in the international arena, since costs would tend to decrease. It was seen as important, however, to be in a position to capitalise on opportunities as they arose - replacing, for example, goods previously supplied by farmers in southern Europe. Ireland should exploit goods like fresh meat, cheese, yoghurts, fruit and vegetables whose price varies little with increasing supply, rather than the staples like cereals and potatoes. The increased demand for more 'healthy/naturally produced' foods should present opportunities in the market place. Demand factors at work outside the country might well be of vital importance to Ireland.

Forestry

The Forestry Study was carried out by the Research and Development Division of Coillte. It was noted that any change which occurs in the Irish climate over the next forty years or so will influence not only the tree crops being planted today, but also many of the trees and forests already growing on this island. Irish forestry, moreover, is currently going through a major transition, and any future effects brought about by a changing climate may well be superimposed on even greater changes due to socio-economic causes.

Higher temperatures and carbon dioxide levels were likely to alter the productivity of our forests both directly and indirectly. The influences at work would include changes in the rates of photosynthesis, in water-use efficiency, in photosynthate allocation, and damage from injurious agencies.

Of the variables likely to affect growth, temperature was thought to be the most important, and would have an enhancing effect. Higher levels of carbon dioxide would also act positively, but the impact would be less significant. Forests were likely to react to the new scenario in one of two ways:

- *Increased* productivity where soil water was not a limiting factor, and where nitrogen availability was enhanced; or
- *Decreased* productivity where the availability of water and nitrogen became limiting factors.

Since the availability of both water and nitrogen are soil related, the regional effects of climate change on forestry would depend, *inter alia*, on the distribution of various soils over the national landscape. Western peatland plantations in areas of high rainfall and low evaporation would benefit, provided the soils were properly drained; with inadequate drainage, however, reduced aeration might prevent trees from taking full advantage of the higher temperatures - particularly if winter rainfall increased as suggested. On mineral sites in the drier parts of the country - the east and south-east - moisture deficits might become evident and have an effect.

Exposure to low temperatures for a period during the winter is necessary for many species of trees if they are to bud normally the following spring - the so-called 'chilling' requirement. If mean daily minimum temperatures were to rise substantially, this requirement might not be met, and some species might cease to thrive. Frost damage, however, insofar as it affects certain sensitive species, would obviously decrease in the new scenario.

The effect of climate change on tree diseases is uncertain. Some diseases could be expected to decline as they are favoured by cool, moist weather (e.g. white pine rust), whereas others liking dry, warm conditions (e.g. powdery mildew on oak) would become more common. Any future threat to Irish forests as a result of climate change was more likely to come from insects than from diseases; climatic warming could affect the movement within Europe and beyond of more exotic pests than those which currently affect our trees.

Seed production was seen as likely to be enhanced by climatic warming. The choice of trees to plant, however, might need to be re-assessed in the light of change, as would silvicultural policies in general, later in the life of a crop; stands might have to have their rotation length changed, or might require greater or lesser inputs in terms of fertilization, pest management, and so on. It might also be important, in a broader context, to assess the effect of possible changes in climate on the productivity of various agricultural crops with which forestry may compete for available land.

The Green Mantle

The study on the Green Mantle was prepared by a team from Trinity College, Dublin, led by Dr David Jeffrey of the Botany Department. They saw the green mantle environment being influenced directly by the combined factors of increasing carbon dioxide, increasing average annual temperature, and changing patterns of precipitation. Increased plant production, higher standing biomass, and an increased ratio of roots to shoots might be expected in a new scenario. There is also evidence that vegetation produced at elevated CO_2 levels decomposes more slowly than that produced at present-day concentrations. Significant increases in the amount of soil organic matter could therefore be expected as CO_2 levels rise; this would affect soil nutrient and water storage capacity.

The assumed increases in temperature were unlikely to be sufficient to directly alter the ability to survive of any of the species at present common in Ireland. They would, however, respond in their individual ways to the new environment, each having an optimum range of temperature for maximum growth, and this might have an effect on their competitive ability. The continued growth of many species throughout the winter could be expected, and the earlier seasonal development and flowering of some plants. The effect of climate change on woodlands would probably be beneficial.

Peatlands, however, would suffer serious damage if the climate were to change along the lines envisaged. The key process in maintaining peatlands in a satisfactory condition is the inhibition of decomposition. In the new scenario increased summer soil water deficits would lead to accelerated decay rates of peat, by increasing the depth of the acrotelm. Moreover, the rate of addition of recalcitrant plant debris to peat formation would decrease; coupled with the accelerated decay rates, this would lead to a net carbon loss from bogs, and a gradual shift from a wet sphagnum-dominated system to a drier dwarf-shrub or grass-dominated heath. It was suggested that the deterioration of peatlands might be the most deleterious outcome of climatic change in Ireland, and emphasised that the phenomenon is without a technical remedy.

Sand dunes, dependent on vegetation for their stability, would benefit from the higher average temperatures, and from increased nitrogen addition to the coastal environment from terrestrial sources. Storm incidence, however, would also be important, since dune growth seawards is a balance between summer accumulation of sand and winter 'comb-down' by waves.

A change in climate would result directly in a corresponding change in the distribution patterns of many insect species. Bird life, on the other hand, would be indirectly, rather than directly, affected - through the abundance of their food supply locally, or by changes in other parts of their range; many of our species are either partial or total migrants. Estuarine ecosystems, too, would be likely to be modified by several consequences of climatic change, although in general the impact on coastlands and estuaries would result more from changes to mean sea level than from climatic change *per se*. Any reduction of the intertidal area would have a serious effect on estuarine winter bird populations and on passage migrants; the impact would be greatest on the smaller waders of the eastern and southern coasts of Ireland.

Hydrology and Freshwater Resources

Dr Con Cunnane and his colleagues at University College, Galway, were entrusted with the study in Hydrology and Freshwater Resources. They envisaged that under the new scenario, the average winter water level in rivers, lakes and in the soil would be higher than at present. Serious flooding would occur more frequently: in areas which currently experience regular flooding, it would be of greater severity and duration, while some areas, currently flood-free, might experience flooding on an occasional basis.

Climate change might also alter the hydraulic characteristics of some catchments, and hence their response to heavy rainfall. Warmer, wetter weather, for example, might lead to greater vegetative cover, which would tend to retard the catchment response to rainfall; floodpeaks would therefore be delayed and reduced somewhat in magnitude. Warmer, drier summers, on the other hand, might result in less vegetation towards autumn, so that a catchment would respond more rapidly to rainfall - perhaps by producing flash floods; flood peaks would tend to follow more quickly after rainfall, and to be greater in magnitude.

Climate change would bring new challenges in the husbanding of our freshwater resources. Lower average summer flow rates in river catchments would result in less recharge for reservoirs during the summer period. Water shortages would occur regularly and the duration of low water levels would be longer than at present. There would be increased losses due to evaporation from reservoirs, and warmer drier weather would bring increased demands for fresh water. Higher rainfall intensities in winter would result in increased soil and river-bank erosion, ultimately causing increased siltation in lakes and reservoirs and reducing their storage capacity. These developments would necessitate the increased exploitation of Ireland's extensive groundwater resources.

Water quality in some areas could be expected to deteriorate periodically in the new situation, due to a combination of summer drought and higher water temperatures. This would be particularly so in cases where sewage systems discharge into fresh water. By contrast, however, drier warmer summers with higher soil moisture deficits would decrease the risk of surface and groundwater contamination from agricultural waste; since agricultural waste has a much greater potential to pollute than domestic effluent in rural areas, this aspect of water pollution might be improved by a change in climate.

Sea Level

The study on the impact on our coasts of changes in Mean Sea Level was carried out by the late Dr Bill Carter of the New University of Ulster at Coleraine.

He emphasised that the lack of confident predictions about future storminess in the vicinity of Ireland was a major hindrance in assessing risk. It was estimated, however, that some 176,000 hectares (176 square kilometres) or 2.5% of the Republic of Ireland were at risk from sea-level rise; this included areas likely to be eroded, flooded, engulfed, or subject to environmental change. The largest proportion of this land is in the west of Ireland, but the most vulnerable areas were seen to be in the east, where much of the shoreline has been developed for one purpose or another.

The risks might be divided into two. Firstly there was the possibility of the gradual inundation of low-lying lands, through a combination of rises in the groundwater table and occasional marine incursions. The areas most sensitive to this are in estuaries (e.g Malahide, Wexford Harbour, Shannon, and Lough Swilly), bays (e.g. Dundalk, Dublin, Dingle, Tralee, Sligo etc.) or behind barriers (e.g between Wicklow and Bannow in the south-east, and north of Fenit in Co. Kerry). Some of the most vulnerable lands are reclaimed tracts, which would require considerable protection and de-watering if they were to remain agriculturally viable.

Secondly, there was the far more widespread problem of the impact of storms, which, given appropriate coincidences of high tides and wind directions can cause considerable damage. Even under conditions of slight sea-level rise the statistical recurrence interval of extreme sea levels, based on present water levels, falls dramatically; an event which in 1990 might be expected to occur once every 100 years might by the year 2030, given the projected changes, be expected to occur every 5 years.

The east coast, although it experiences lower wave energy levels than the west, was seen as having more potential for shoreline erosion. Very approximate calculations suggested that increased erosion of 15-20% per year might result in the new scenario, although the complex control systems that govern erosion might reduce this figure.

The cost of 'protecting' all vulnerable segments of the Irish coastline by building sea-defenses was estimated by Dr Carter, very roughly, at about IR£270,000M - clearly a figure that could not be contemplated. Other strategies, including abandonment of land, stronger planning control in coastal areas, fiscal or financial penalties for coastal developers, or softer non-engineering protective measurements, would need to be considered.

Fisheries

Aquafact International Services of Newcastle, Co. Galway, undertook the study on Fisheries. They pointed out that Irish coastal waters tend to represent a transition zone, with many species at or near their southern or northern extremes. A general warming of the coastal waters, should it occur, would probably result in shifts in the range and distribution of commercial species, and also changes in their migration patterns and in predator/prey relationships.

Some northern species such as salmon, cod, haddock, herring, and mackerel would be likely to show altered distributional and migrational patterns in such circumstances, while more southern species such as pilchard and hake might become more abundant. Some flatfish species, like plaice, might also increase. Lobsters could be expected to thrive, since higher temperatures tend to extend their active feeding periods and to provide better conditions for the survival of lobster larvae.

Other factors, such as present fishing stresses and overall water quality, were also seen as relevant. In the event of a change in the marine environment, it was envisaged that species which are already stressed would be more severely and dramatically affected by the new scenario than others. New fishing grounds and stocks, and the possibility of catching previously unexploited fish (e.g. tuna), would need to be investigated.

The changes envisaged in precipitation levels might have important adverse consequences for the life cycle of anadromous species - those which migrate from freshwater and feed in the oceans, such as salmon and sea trout. Increased winter precipitation, combined with such transient effects as increased westerly storms and siltation, might result in the disruption of the gravel beds or redds used for spawning; it might also bring about excessive leaching from Carboniferous forest soils, causing increased acidity in rivers and lakes, which might be detrimental to salmonid parr. A reduction in summer rainfall, on the other hand, might cause migration difficulties for smolts; moreover, a rise in average temperature, with consequent reduction in the dissolved oxygen content of rivers and lakes, might affect the viability of salmonid fish, since they have high oxygen requirements.

The consequences for aquaculture would be varied. In the case of the salmonid species, higher sea temperatures might well cause an increase in sea or salmon louse infection and a potential risk of an increased rate of spread of various fish diseases. Increased irradiance in summer might also cause an increase in the incidence of sunburn, as fish come closer to the surface to avail of higher oxygen supplies. Salmon smolt production in freshwater hatcheries might show decreased productivity in a regime of higher average water temperature. Rainbow trout would experience problems similar to those faced by salmon.

In the case of shell-fish, the consequences of climatic change might be expected to be generally beneficial. Growth in cold-blooded organisms is accelerated by higher temperatures, and therefore a warming of the sea water should have a slightly favourable or neutral effect on their culture (although there would be a few exceptions). The native oyster, for example, is at its northern limit of distribution in Irish waters, and a potential rise in temperature might enhance its growth and its potential for spawning in the wild. The cultivation of mussels should also be enhanced, and the production rates of clams and scallops would also improve under the suggested new scenarios. The higher temperatures might allow the introduction, or increased production, of alternative species such as the abalone or the sea urchin.

Some adverse consequences were also foreseen. Higher temperatures would increase the risk of the more rapid spread of potentially damaging diseases; changes in precipitation patterns might increase the sediment load in river run-off, which might in turn disrupt beds and reduce yields; increased siltation rates might also clog the feeding apparatus of adult animals and lead to fatalities.

Monitoring of red tides would become more important, as would the need for more efficient toxin detection techniques.

Reference:

McWilliams, B.E. (Ed.), 1991. *Climate Change : studies on the implications for Ireland,* Department of the Environment, Dublin, published by the Stationery Office, 193pp.

PREDICTED CLIMATE CHANGE AND ELECTRICITY GENERATION

Owen Wilson
Electricity Supply Board

The instructions on writing this paper were to describe what current policy is in relation to climate resources, which for the electricity sector effectively means policy in relation to CO_2 emissions control, and then to suggest what the policy should be. An attempt at answering these questions is given below. However, to provide some background I would first like to note some points to bear in mind in examining the policy questions related to a potential enhanced greenhouse effect from an electricity utility's perspective. These include:

- Policy in relation to electricity is a subset of energy policy, the scope of action of which is limited by the requirements of other national priorities, including social, economic and environmental.
- Electricity production in the European Union (EU) accounts for approximately 30% of total primary energy requirement and roughly the same percentage of carbon dioxide (CO_2) emissions (for Ireland, 33% and 35% respectively).
- Electricity's share of the final energy demand is increasing in all countries and will continue to do so for the foreseeable future.
- Nuclear power provides a number of national utilities with a possibility to stabilise their emissions of CO_2.
- Thermal power stations, as highly visible point sources of emissions, are the focus of a significant amount of policy-maker's attention.

The possibility of future global warming is a key concern of the electricity sector world-wide. This is a consequence of the potential future impacts such warming could have in relation on electricity demand, on power generation and on the management of resources. The sector is equally concerned about the impact of policy responses and strategies being proposed to address this potential threat on the operations of the industry. The problem facing policy-makers, having agreed to control CO_2 (and other greenhouse gases) emissions, is, in simplistic terms, that growth in electricity and energy demand is a function of both population growth and economic development. Because limited success is being achieved in controlling population growth, because economic development is viewed as essential in eliminating global poverty and inequality and because fossil fuels will continue as the main form of energy and will not be a limiting factor on development for the next century, emissions of CO_2 are inevitably set to rise.

To counter this future growth in emissions will require an improvement in the carbon intensity [GNP($)/Carbon(t)] of energy supplies and continued and rapid advances in the energy intensity [GNP($)/Joule] of global economies. Measures which will impact on these two objectives include improvements in the conservation of energy and the efficiency of energy supply and use, increased penetration of renewable energy sources and, where socially and politically acceptable, nuclear power, the removal of structural issues inhibiting the rational use of energy and, most critically, the availability of finance to undertake the necessary investments. However, even with sufficient capital available, introducing major changes to the energy supply structures will take time, which means that rapid reductions in CO_2 emissions are not achievable in the short term without major social and economic dislocation.

In the interim, a number of strategies have been developed at national and international levels to address the control of CO_2 and other greenhouse gas emissions. These measures broadly include the following: financial instruments, voluntary self-commitments, command and control legislation (i.e. for CFCs) setting standard norms, public education campaigns and scientific and technological research, development and demonstration.

What is policy in relation to CO_2?

Irish policy in relation to CO_2 emissions control is defined within EU and UN contexts. Essentially these require that we contribute to the EU objective of aiming to stabilise Union emissions at 1990 levels by 2000. The need for equitable burden sharing by member states in achieving this objective has facilitated the government in setting the national objective at limiting the growth in CO_2 emissions to a net 11% after allowance is made for an increase in carbon sequestration.

The application of national policy on CO_2 emissions control to the electricity sector is subject to a number of constraints. In essence these relate to the elements of conflict which arise between the priorities of economic, social and energy policies i.e. the desire to maintain regionally balanced, strong economic growth and to continue to utilise indigenous energy resources. The most important for ESB is energy policy, which as applied to the electricity sector, has three key priorities: security of supply, minimum cost to the consumer and due regard to the needs of the environment. These priorities relate back to economic and social policy, in that competitiveness depends, in part, on energy costs and security of supply considerations support the development of indigenous fuel resources. These factors also impinge on environmental priorities. On the one hand, specific policy responses to CO_2 control which generate costs without compensating savings have a low priority [CO_2 scrubbing from flue gas] while on the other, control measures which impinge negatively on social policy areas also have limited support irrespective of cost implications [fuel switching away from peat, which is a high CO_2 emitter].

The principles on which specific national policy for the electricity sector is based relate to the need to take precautionary action and the desire to ensure that the polluter pays. How this is to be implemented is outlined in the recent strategy document issued by the Department of the Environment (McWilliams, 1991).

The document recognises the essential role of electricity in our technology-based society and its importance in providing for economic growth. ESB is required to:

- Manage load growth towards an annual average rate of 3% p.a. to 2000 and 2.5% p.a. thereafter, without constricting economic growth.
- Manage the fuel mix to minimise emissions of CO_2, consistent with the need to maintain a reasonable price for electricity.
- Continue to expand existing demand side management programmes in all market segments.
- Continue to support supply side efficiency improvement programmes.
- Utilise integrated resource planning, comparing investments in demand management with new plant to achieve a 'least cost' solution.
- Promote the use of CHP at commercial and industrial locations.
- Continue to upgrade to distribution network to minimise power losses.

ESB is already implementing the above measures. ESB has also introduced on a pilot basis a Third Party Finance scheme to provide certain customers with the opportunity to make energy efficiency and conservation improvements. All the above are being undertaken because they are economic in their own right in addition to controlling emissions. In addition and in accordance with a specific commitment in the Programme for Partnership government, ESB will source a further 75MW of alternative energy capacity by 1997.

What should policy be?

Ideally it should be rational in its application of the precautionary principle, while attempting to integrate the conflicting requirements of other policy priorities. In this context it has succeeded to date. After that there are no simple or unique solutions and a range of actions are required. This policy must be applied so that least cost options are exhausted before more expensive measures are adopted. Where measures are being implemented, government initiatives should aim to optimise industry and consumer responses.

The key strategic priority must be to reduce the degree of uncertainty concerning the predictions of an enhanced greenhouse effect. Major uncertainties remain in relation to both the science of the processes involved and the climate models. Therefore, scientific and modelling research must be continued so as to improve the confidence level of the predictions. Research is also required in relation to the economic costs associated with both control and adaptation measures. In this context, a recent U.S. OTA study indicated that the economic costs of adaptation could be substantially lower than those required to stabilise CO_2 emissions. Similarly, at a recent conference in Holland it was reported that measures taken to control non-CO_2 greenhouse gases, responsible for 50% of the predicted effect, could be taken at much lower costs than CO_2 stabilisation.

While uncertainty remains with regard to the science underlying the hypothesis and the model predictions, there is also uncertainty regarding future technological developments in the generation, supply and use of electricity. It is important that the maximum flexibility of the electricity sector to adapt to unforeseen advances in technology or changes to global warming predictions is maintained. This has a number of consequences. It implies that research and development must continue to be supported on a wide range of future generation options and technologies. It also suggests that the real cost of energy must be applied to the consumer and that transparency must be introduced into energy and electricity pricing. Most fundamentally, however, it requires that measures supporting the rational use of energy across all sectors be adopted.

The unique role of electricity should also be recognised. While significant efforts have been made at improving energy intensity and decoupling energy demand from economic growth, a strong relationship continues to exist, even in highly developed economies, between economic development and electricity demand. The potential of electricity for solving part of the ghg emissions problem has been proven over the past decade. Measures must not now be put in place which could inhibit electricity from making a similar contribution in the future. For the ESB this implies that electricity use, and its consequential CO_2 emissions, must be allowed grow to accommodate situations where electricity is substituted for other forms of energy and thus optimises the overall environmental benefit.

Issues of joint implementation of emission reduction objectives need clarification. It must be possible within a national economy to account readily for complementary measures leading to reductions in ghg emissions e.g. the substitution of electricity for a more polluting energy source. In addition, trans-national measures, such as support for energy efficiency improvements or the expansion of carbon sinks in developing countries, can often provide cost-effective opportunities for greenhouse gas emissions reductions. An effective mechanism for accrediting such savings must be developed and impediments to undertaking joint actions on a global scale should be avoided.

Finally, should the predictions be confirmed, the question as to the availability of capital to undertake the major structural reorganisation of the energy supply and, in particular, the electricity supply systems must be considered.

Reference

McWilliams, B.E. (Ed.), 1991. *Climate Change : studies on the implications for Ireland,* Department of the Environment, Dublin, published by the Stationery Office, 193pp.

WIND ENERGY IN IRELAND – A SURVEY OF RELEVANT ISSUES

Rick Watson
Department of Electronic and Electrical Engineering
University College Dublin

1. Introduction

Wind energy can be regarded as a climate resource. A resource that we have an abundance of in Ireland and which until recently has been largely untapped. It is an indigenous resource which can meet a significant proportion of our electricity needs. A resource which can lead to more balanced regional development. A resource to help limit SO_2 and NO_x emissions. A resource for which the economics are in the right ballpark and improving. A resource which in the context of possible global warming presents us with an opportunity to limit CO_2 emissions. A resource we should support by appropriate measures. This paper presents a survey of the issues relating to wind energy in Ireland.

2. Current Status of Wind Energy in Europe

Wind energy technology has made great strides over the last decade. Reliable, efficient wind turbines are now available up to ratings of about 500 kW. The wind energy resource is being developed in many European Countries. The Eurowin database registered 6078 wind turbines with a total capacity of 982.1 MW in the EU countries at the end of September 1993 (Klein, 1993). It is reckoned there is another 200 MW of capacity which is not registered with the database giving a total of nearly 1200 MW. The capacity is distributed among the different countries as shown in Figure 1. Denmark is clearly the leader in terms of installed capacity but there have been substantial developments in the Netherlands and Germany and more recently in Spain and the UK. The Eurowin database production figures for 1992 (the last year for which data are available) are 896.5 GWh from an installed capacity of 597 MW which reported to the database for that period, giving an overall capacity factor of about 17%. There is one windfarm operating in Ireland consisting of 21 wind turbines of total capacity 6.45 MW located at Bellacorick, Co. Mayo. This project was grant-aided at 55% of the project cost under the Valoren programme. Its first year performance has been up to expectations with a capacity factor of about 29%. Two further windfarms of capacity 1.2MW and 5MW are planned for Arigna, Co. Roscommon and Castleisland, Co. Kerry and have been approved for support under the Thermie programme. The 5MW windfarm has been refused planning permission by Kerry County Council and is being appealed to An Bord Pleanála.

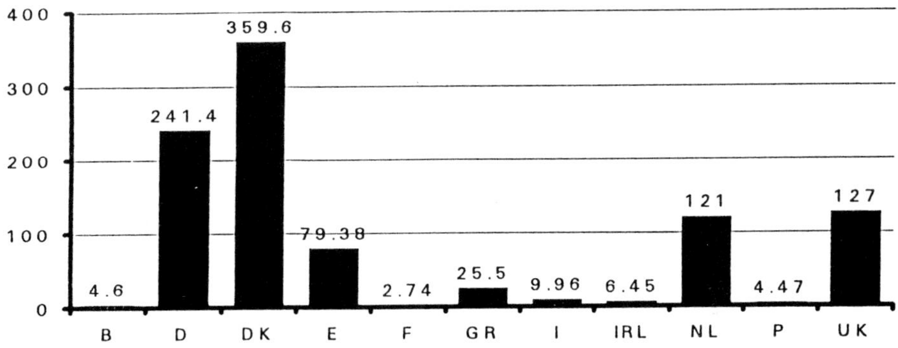

Figure 1. Installed wind energy capacity [MW] in the EU countries as registered in the Eurowin database up to September 1993

3. Wind Energy Resource and its Development Implications

The wind resource in Ireland is enormous. The raw resource potential could meet not just our electricity needs but all our energy needs many times over. In Table 1 below an estimate is given of the number of turbines required, the required installed capacity and the area of land (in hectares and in % of the Irish land area) over which the machines would be spread to meet 10%, 20% and 100% of our current electricity demand. The calculations are based on the use of 500 kW, 40 m diameter wind turbines at 8 rotor diameter spacing grouped in 8MW windfarms operating at an annual capacity factor of 25%.

Electricity production	No. wind turbines	Installed capacity [MW]	Spread over land area [ha]	Percent of land area [ha]
1.5TWh (10%)	1,376	688	22,000	0.31%
3TWh (20%)	2,752	1,376	44,000	0.62%
15TWh (100%)	13,712	6,856	220,000	3.13%

Table 1. Number of wind turbines, installed capacity and land required to meet different percentages of current electricity consumption

This land use can be contrasted with that of another renewable resource of which there is an abundance in Ireland, i.e. biomass or more particularly short rotation forestry (SRF). In Table 2 the land requirements to produce the same proportions of our electricity use are estimated on the assumptions of a four year cycle SRF operation where harvests of 12 t/ha are achieved, the biomass produced has a calorific value of 4.5 kWh/kg and is burned in 30% efficient power stations. It is clear that wind requires a very much smaller area of land and has the added advantage that the land is not sterilised in that the land between the machines can be put to other uses e.g. animal grazing. In fact, in a typical case about 1% of the land area over which a windfarm is spread is actually used for access roads, parking and foundations.

production	land area [ha]	percent of Irish land area
1.5 TWh (10%)	370,400	5.27%
3 TWh (20%)	740,800	10.54%
15 TWh (100%)	3,704,000	52.7%

Table 2. Land use of biomass for different electricity productions

There are however limitations on how much wind generation could be absorbed into ESB's power system. These limitations are on two levels and for two distinct reasons. On the level of the national transmission grid ESB have identified in a recent study a technical limit of 7.5% of electricity demand before frequency regulation problems would be encountered (O'Dwyer, 1991). This limit is particularly low because ESB operate an island power system - there are no interconnections to neighbouring power systems - which means it has to carry all its own spinning reserve. The figure should be improved upon if the interconnector to the NIE system is re-established and if a new interconnector is established to the system in Britain, which seems likely given that it would probably be supported under the EU single energy market policy. If further pumped storage capacity were to be added to the ESB system in the future, and with likely developments in power system control, is seems likely that this limit could be pushed up to the 15% to 20% range in the medium to long term. Looking even further into the future, and with the possibility of the widespread use of the electric car coupled with flexible control of the recharging load, it is possible that higher penetrations of fluctuating resources like wind could be achieved.

The second limitations is on the amount of additional generation capacity which can be embedded into local distribution networks. In Ireland the population density is low, electricity usage is low and the distribution networks in rural areas are 'electrically' weak, particularly so in the remote exposed areas which would be ideal from a wind resource point of view. This problem can be overcome but it entails the building of new networks or reinforcement of existing ones, both of which involve substantial cost penalties which would fall on the windfarm developer.

There are many positive implications for regional development if the wind energy resource is to be developed. The first has already been referred to, i.e. a likely improvement in electricity distribution networks which could be of benefit to existing or more particularly new electricity consumers. There would also be an improvement required in the local road network. New roads for access to the wind turbines for cranes during construction and later during maintenance would be required (typical total nacelle weight for 500 kW wind turbines would be in the range 10 to 15 t). These roads could then be put to other use, e.g. access to land for farming. It might also be the case that some improvements of existing roads would be required.

Initially, wind turbines would be imported from one of the major supply countries (Denmark, US, Germany or Holland) but if a sufficient market developed here then there would be no reason why a licencing agreement to manufacture here in Ireland would not be justified. Initially then jobs would be created temporarily during construction, in consulting and services and in operation and maintenance. Later there could be jobs in manufacturing.

There would also be direct benefits to landowners on whose land the wind machines would be placed in the form of a rent or outright purchase. So windfarming could become an alternative source of income from agricultural land. It is very important that there also be some financial benefit to the local community as a whole. This could take the form of joint ventures with local co-operatives, or the possibility of local interests investing in the projects or in the form of a local tax set as a percentage of the income of the project. This would have the effect of dispersing some of the economic benefits throughout the local economy and not allowing all the benefit to accrue to outside or foreign investors.

The benefits on a national level are firstly that energy imports would be displaced by developing an indigenous resource, secondly there would be a strategic benefit in diversification of the energy supply mix, and thirdly there would be benefits in helping to meet EU agreed targets on emissions of SO_2, NO_x and CO_2 to the atmosphere.

4. Environment

The harnessing of all energy sources involves environmental impacts. The exploitation of the wind energy resource is no exception. The impacts may be direct or indirect, visible or invisible. The environmental impacts of wind energy are visible (you see what you get) but the substantial environmental benefits when compared with conventional energy sources like fossil fuels and nuclear are quite clear as summarised in Table 3.

Environmental Impact	Fossil Fuels	Nuclear	Wind
SO_2, NO_X emissions	Yes	Indirectly in uranium mining and processing	None
CO_2 emissions	Yes	Indirectly in uranium mining and processing	None
Waste heat emissions	Yes	Yes	None
Pollution during fuel transportation and processing	Yes (e.g. oil spillage)	Yes (e.g. mining)	None
Low level radiation emissions	None	Yes	None
Risk of accident involving major release of radioactive products	None	Yes	None
Waste products	Yes (e.g. ash)	Yes (supervised storage for thousands of years)	None
Decommissioning Cost	Low	Unknown	Low

Table 3. Comparison of environmental impacts of wind and conventional energy sources

These environmental benefits have to be weighed against the environmental impacts of the development of the wind energy resource. These are: visual impact, noise emissions, electromagnetic interference, land use implications and effects on flora and fauna. In order to maximise the energy production of wind turbines they should be sited in exposed locations in regions of high mean windspeeds. Prime locations would be exposed, smooth ridges and hills where visual impacts on the landscape would be an issue. This would also apply to the power lines which would be required to transmit the output of the windfarms to the grid. The assessment of visual impact is highly subjective. It is essential that local communities be involved in real decision making at the planning stage. It has been found that there is strong correlation between local benefits to the economy and acceptance of any real or perceived visual intrusion. Guidance for planning authorities is an essential requirement particularly in the current situation in Ireland where the planning authorities are working in the dark on this issue.

There are two sources of noise from a wind turbine, the swishing aerodynamic noise of the blades and the machinery noise of the gearbox and drive train in the nacelle of the machine. The machinery noise can be reduced by good design and appropriate noise insulation of the nacelle. The aerodynamic noise is really only a problem at windspeeds below 8 m/s, as above this windspeed the ambient noise increases (wind in trees etc.) and blankets the noise of the wind turbine. Noise from wind turbines should not be a nuisance if careful design is adopted and the standard of 500 m to the nearest dwelling is accepted.

Interference with TV reception is not usually a problem but can easily be solved. Microwave links need to be avoided. As has been seen already wind turbines need to be spread over a large land area. However the land is not sterilised but is available for normal agricultural use. The effect on flora is minimal. The normal habitat is disturbed during the construction but returns to normal following completion of the construction. The effect on birds is also minimal. Various studies have indicated bird deaths associated with a line of wind turbines would not be any greater than for a high voltage power line of equivalent dimensions (Garrard, 1991).

In general it is clear that the environmental benefits of wind energy outweigh the concerns. The main problem is that the benefits of wind energy might be regarded as global or national, whereas the concerns are local, and due to the diffuse nature of the wind would directly affect a larger number of people than the conventional generating station, which is a large point source of environmental impact which directly affects a smaller number of people, and whose indirect adverse effects are felt nationally or globally.

5. Economics

Many factors affect the economics of wind energy projects. The most important are capital cost, energy production (which depends primarily on available windspeeds), operation and maintenance costs, interest rates and lifetime. The graph in Figure 2 below illustrates the effect on the production cost of electricity of varying one of five parameters on a percentage basis around a base case value while keeping the other parameters fixed. The five parameters chosen and their base case values and ranges of variation are given in Table 4. The base values chosen are considered realistic for the present day e.g. capacity factor 30%, capital cost £900/kW, interest rate 10%, lifetime 10 years (the typical design machine lifetime is usually 20 years but 10 years would also correspond to a typical period from a financing point of view), O & M rate 5%. A mechanical availability of 95% is also assumed.

Parameter	Base Case Value	Range of Variation
Capacity factor	30%	15% to 45%
Lifetime	10 years	5 years to 15 years
O&M rate	5% per annum (of capital cost)	2.5% to 7.5%
Specific capital cost	£900/kW	£450/kW to £1350/kW
Interest rate	10%	5% to 15%

Table 4. Assumed base case values and ranges of parameters for variations in range ±50% about base value

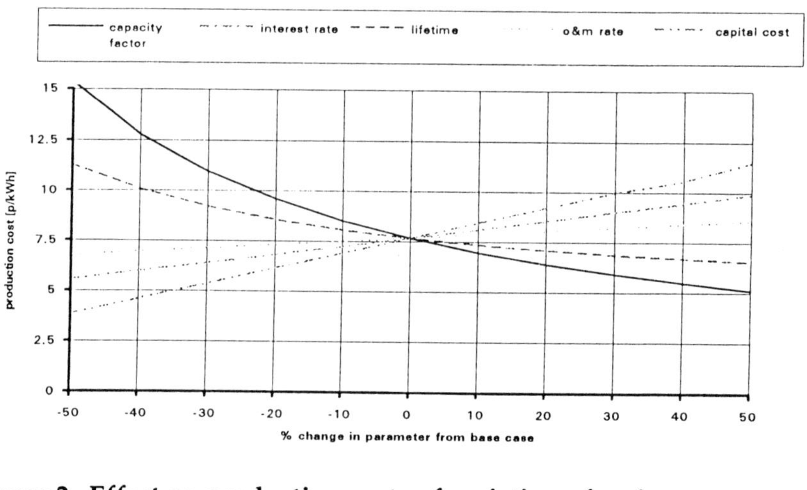

Figure 2. Effect on production costs of variation of main parameters affecting the economics

It is clear that the capacity factor has a very strong influence on production costs. The capacity factor is determined mainly by the wind speeds available at the site. So good wind sites are essential for economically viable projects.

The current buyback rates available from the ESB and the projected price increase for non fuel producers are listed in Table 5 below for supplies at 10kV, 20 kV and 38 kV up to 25 GWh per annum.

	Day (mon-fri) 8am to 9 p.m.	Night and weekend	Average
Present	6.08 p/k Wh	2.40 p/k Wh	3.82 p/k Wh
From 01.06.94	6.26 p/k Wh	2.47 p/k Wh	3.93 p/k Wh
From 01.06.95	6.45 p/k Wh	2.54 p/k Wh	4.05 p/k Wh

Table 5. Buyback rates available from ESB

It is clear that some incentive is required to stimulate investment in wind energy. This could take the form of a capital subsidy, improved buyback rates or tax concessions. These measures can be justified on the basis of taking into account the social costs of conventional electricity generation, which are not normally included in economic analyses when making comparison with wind energy as described in Hohmeyer (1988).

In October 1993 the Minister of State for Energy announced an Alternative Energy Requirement which will ensure that 75 MW of alternative energy capacity (including renewables and CHP) will be installed on the ESB system before 1997. The details of the scheme are being drawn up by the Department of Energy and ESB and it is hoped they will be finalised within the next few months.

6. Conclusions

There is no doubt that wind energy technology has made major strides in the last decade. It is in the right ballpark from an economic point of view and will likely improve its position vis-a-vis the conventional technologies. Incentives are required to stimulate the market. These incentives can be justified if external costs of conventional electricity are included.

The following actions have been put forward as being needed to help Ireland exploit its vast wind energy resources (Watson, 1988, 1991):

(1) Detailed wind resource assessment studies.

(2) Detailed windfarm feasibility studies.

(3) Detailed system integration studies.

(4) Design, installation and operation of pilot windfarms by ESB to gain operational and maintenance experience.

(5) Private windfarm development to be encouraged by better buyback rates (to reflect the full social costs) and contracts.

(6) Irish industry involvement in joint ventures with existing wind turbine manufacturers.

Much still needs to be done but there has been considerable progress. Wind resource assessment studies are being carried out by the author in UCD with EU funding (Watson, 1993), by ESB and by some private developers. More extensive studies are required. The ESB have carried out a wind penetration study referred to earlier but again more detailed studies on the national grid level and also on the local distribution level are required. The first windfarm is installed and operating well. Others are planned. Details of the Alternative Energy Requirement are expected shortly from the Department of Energy.

Some further actions that could be added to the list are firstly the need to inform the planners on the issues involved and to set out guidance principles for them, and secondly to inform the banking/insurance sector on the advanced state of the wind turbine industry. Finally a useful model for development would appear to be a joint venture between local interests and utilities (Moerup-Pedersen and Pedersen, 1992) which when applied to the Irish situation would involve local interests, ESB and private investors. The involvement of ESB would help ensure quality, reliability and safety standards, that of local interests would help ensure public acceptance and that of the private investors would ensure the introduction of diversity and an element of competition into the electricity supply market. It is clear that Ireland is poised on the threshold of a major increase in wind energy utilisation.

References

Garrad, A., 1991. *Wind Energy in Europe - Time for Action,* European Wind Energy Association.

Hohmeyer, O., 1988. *Social Costs of Energy Consumption,* Springer-Verlag, Heidelberg.

Klein, H-P., 1993. Actual Survey on European Wind Energy Technology from the EUROWIN Database, Seminar on Wind Energy in Southern Europe, Cadiz, Spain.

Moerup-Pedersen, V. and Pedersen, S., 1992. *Wind Farm Projects as Joint Ventures Between a Danish Utility and Private Cooperatives, The Potential of Wind Farms,* European Wind Energy Association, Denmark.

O'Dwyer, E., 1991. Wind Energy from a Power System Perspective, Solar Energy Society of Ireland Conference, UCG.

Watson, R., 1991. Wind Energy, In Feehan, J. (Ed.), *Environment and Development in Ireland,* 202 - 207.

Watson, R., 1988. Wind Energy in the Republic of Ireland, British Wind Energy Association, Wind Power Generation Conference, Queen's University Belfast.

Watson, R., 1993. *Wind Measurement and Modelling in Ireland,* ECWEC, Travemuende, Germany.

PEATLANDS, CARBON CYCLING AND POLICY ISSUES

Charles Shier
Bord na Móna, Leabeg
Tullamore, Co. Offaly

1. Introduction

Peat deposits cover 400 million hectares, or almost 3 per cent of the Earth's land area. These deposits constitute a significant carbon pool, containing an estimated 460 gigatonnes (460 x 10^9t) of carbon. This peatland pool equates to 60 per cent of the total carbon store in all vegetation types, and contains over three times the amount of carbon sequestered in the Earth's tropical rainforests.

Actively growing peatlands have a positive net carbon balance, and act as effective carbon sinks. The actual rate of accumulation depends on deposit age, latitude and hydrological and trophic status. Disturbed peatlands, where natural processes have been disrupted through anthropogenic intervention, may have positive, neutral or negative carbon balances, depending on the specific utilisation of the peatland resources.

This paper briefly examines the current status of peatlands in the context of prevailing policies and anticipated climatic change. Section 2 discusses the morphology of peatlands and the processes involved in the cycling of carbon. The impacts of peatland drainage and disturbance, and the emissions associated with differing peatland uses, are explored in Section 3. Peatland carbon fluxes, at the global, European and national levels, are quantified in Section 4; while Section 5 describes the historical and recent policies relating to the utilisation of Irish peatlands and examines their applicability in the light of changing environmental objectives.

2. Peat formation, Mire structure and Carbon Exchange

Peat is an organic material that accumulates in surface deposits as a result of the incomplete breakdown of wetland vegetation. While vegetative decomposition occurs rapidly under aerobic conditions, it is arrested, or at least retarded, under conditions of excess moisture and oxygen deficiency. Peat may be found throughout the world, anywhere the natural drainage of rainwater is reduced or impeded. It is, therefore, frequently located in deltas and estuarine systems, along river terraces and in lake and former lake basins. Alternatively, peat formation may arise through climatic induction, where wet and humid climates lead to saturated ground conditions and lower temperatures impede plant decay.

In Ireland, peat deposits are found in the Central Plain, along the western seaboard, and in mountainous regions. The deep, extensive deposits in the Midland counties formed initially in post-glacial lakes, left by the retreating ice some 12,000 years ago. The shallower peat deposits, found along the West Coast and at higher altitudes everywhere, are climatically induced. Their presence is mainly due to increased rainfall or lower temperature, both of which inhibit decompositional processes.

Within a peat-forming ecosystem, or mire, two morphologically distinct layers may be identified: (i) a shallow surface layer that includes the living vegetation and recently dead plant material, known as the **acrotelm**; and (ii) the underlying body of accumulated organic remains, or **catotelm.** The thickness of the acrotelm is determined by the depth to which the water table falls in summer. Within this layer, interstitial spaces have free exchange with the atmosphere, and decomposition is predominantly aerobic. In the permanently saturated catotelm there is limited diffusion of atmospheric oxygen, and the processes of decay are predominantly anaerobic.

As peatlands grow and develop they act as effective sinks for atmospheric carbon, with the rate of carbon accumulation depending on the balance between the sequestering and the liberating processes. Carbon dioxide (CO_2) from the atmosphere is fixed through photosynthesis by the living plants in the surface layer (Fig. 1). As these plants die and decay, some of the fixed carbon is converted back to carbon dioxide in the aerobic acrotelm. Only a proportion of the carbon originally fixed passes through to the anaerobic catotelm. Here the rate of decay is much, much slower, and the mass of organic material gradually builds up. The main products of anaerobic decomposition in the catotelm are methane (CH_4) and carbon dioxide, in roughly equal amounts. Part of this methane is reoxidised, to carbon dioxide, as it passes up through the oxic surface layer.

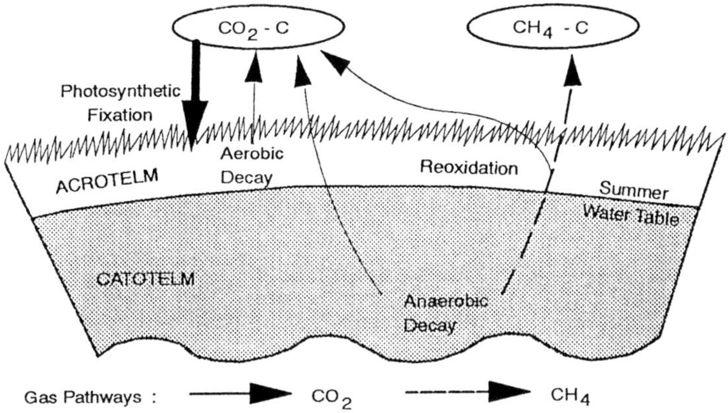

Figure 1: Mire Structure and Carbon Exchange

The net rate at which a peat deposit captures atmospheric carbon changes over time (Fig.2). Carbon fixed through photosynthesis in the acrotelm may be passed to the underlying catotelm at a fairly constant rate. However, as the mass of peat accumulates, the net effect of anaerobic decay in the catotelm becomes increasingly significant. The catotelmic release of carbon gradually increases and eventually approaches the rate at which it is passed from the acrotelm, ultimately resulting in a 'steady state'.

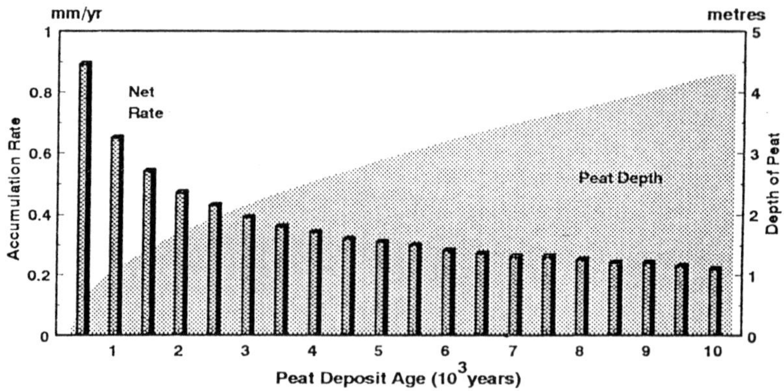

Figure 2: Variation in Rate of Peat Accumulation in Boreal Mires

Superimposed on this natural ecosystem ageing process, primary productivity, and the net carbon accumulation rate, may also be affected by climatic changes over time. Evidence from Southern Finland and from Maine in the United States suggests that underlying mire carbon accumulation rates were more rapid during the early Holocene, slowed about 8,000 years ago, and only increased again in the last two to three millennia. Similar accumulation rate changes have also been found in Eurasian mires. This common pattern is attributed to changes in peat-forming processes in the Hypsithermal, and the subsequent accelerated deterioration of climate in recent millennia (Tolonen *et al.*, 1992).

Regional storage rates vary today from less than 0.2 tonnes of carbon per hectare per annum in the higher latitudes of the boreal zone (Canada, Alaska, Scandinavia and the former USSR), to a maximum of 2.25 tonnes of carbon per hectare per annum in Florida (Armentano and Menges, 1986). In summary, all mires in their natural state exhibit a number of characteristics that have climatic connotations: they sequester carbon from atmospheric carbon dioxide; they convert some of the fixed carbon into methane; and they become gradually less effective in sequestering carbon as they age.

3. The Impact of Peatland Development

The disturbance of peat-forming ecosystems, through drainage and development for forestry, agriculture or peat production, or through other anthropogenic interventions such as the cyclical burning of surface vegetation, has a significant impact on the exchange of radiatively active gases. The magnitude of this impact depends on the exact nature of the intervention, and on the trophic status of the mire. In general, when peatlands are developed the initial drainage activities typically:

* increase the depth to the water table;
* increase the thickness of the oxic layer;
* increase the amount of carbon dioxide released;
* decrease the amount of methane released.

Recent evidence from Finland also indicates that drainage of nutrient rich (minerotrophic) mire types may lead to the release of nitrous oxide (N_2O). Although the amount of nitrous oxide released is very small in comparison to the fluxes of carbon dioxide and methane, nitrous oxide is more efficient as a greenhouse gas, has a long atmospheric residence period, and has also been implicated in the chemical reactions that destroy ozone in the stratosphere (Martikainen *et al.*, 1993).

Agriculture. The magnitude of the increase in carbon emissions, from peatlands reclaimed for agricultural purposes, depends on their latitude and specific end use. Arable cropping, which usually requires deeper drainage, regular cultivation and significant lime and fertiliser inputs, results in much faster oxidation of the organic material than permanent pasture. Carbon release from peatland areas growing crops ranges from 2.2 t C ha^{-1} yr^{-1} in Finland to 21.9 t C ha^{-1} yr^{-1} in Florida. Comparable figures for areas under permanent pasture range from 0.3 to 15.5 t C ha^{-1} yr^{-1} (Armentano and Menges, 1986).

Forestry. The net effect on carbon exchange is more complex in peatland areas that have been reclaimed for forestry. Increased carbon dioxide emissions, caused through lowering of the water table, are offset by the reduced emissions of methane, and by the accumulation of carbon in the trees, wood products, litter and forest soil. Studies on conifer plantations on drained peatlands in Britain indicate that the amount of carbon sequestered by plantations reaches a maximum over 2 to 3 rotations. In the short term, over the next 50 to 200 years, afforestation of peatlands in temperate regions will increase the net carbon capture, provided that the rate of oxidation of the peat substrate remains low (Cannell *et al.*, 1993).

Non-Fuel Peat Production. While the production of peat is frequently associated with the supply of energy, peat also has a considerable number of alternative uses. These include plant growing media, soil improvement, litter for animal bedding, the production of peat coke and activated carbon, and the application of peat in biofiltration and for balneological purposes (peat spas). Although the half-life of the peat material varies considerably, depending on the actual application, all of the peat will ultimately be broken down and the carbon component released. Over the longer term, provided that the rate of production remains stable, and the proportional use doesn't vary significantly, the annual release of carbon will approximate the annual peat extraction rate.

Fuel Peat Production. The utilisation of peat as an energy source results in a much more immediate release of carbon. Energy products, such as milled peat, sod turf or peat briquettes have a relatively short storage life, and combustion processes ensure almost complete reoxidation. In geological terms peat is sometimes regarded as a 'young coal', as some contemporary peat forming communities bear certain similarities to the swamp conditions of the Carboniferous period. However, fuel peat also exhibits some striking disparities with fossil fuels, such as coal, oil or natural gas:

FUEL PEAT	FOSSIL FUELS
* Surface deposit	* Generally within the Earth's crust
* Recent formation	* Formed over geological time scale
* Formed through microbial processes	* Geological/geochemical formation
* Active carbon accretion	* No active carbon sink
* Chemically equivalent to wood	* Higher carbon, lower oxygen content
* Can be restored post-production to an active carbon sink	* Cannot be restored

In Ireland, milled fuel peat extraction rates may range from 55 tonnes per hectare per annum, in a poor production season, to 280 t ha^{-1} yr^{-1} in a year characterised by long periods of stable, anticyclonic conditions during the summer months. A buffer stock management system ensures a more uniform annual offtake of circa 180 tonnes of fuel peat per hectare for combustion purposes, resulting in the release of c. 45 tonnes of carbon per hectare each year.

4. Carbon Fluxes in Peatlands

In a global context, significant areas of the world's peatlands have not yet been fully inventoried, their exact status remains inadequately defined and data relating to carbon accumulation and release are incomplete. Using best estimates of the areas drained for forestry and agricultural purposes, and recent figures for peat production activities (Couch, 1993), global peatland emissions total an estimated 128.5 million tonnes of carbon per annum. To help put this total in perspective, it is compared with the estimated global carbon flux, and the fraction attributed to anthropogenic activities, in Figure 3. Estimates of net global carbon capture by undeveloped peatlands vary considerably. Early estimates ranged between 135 and 400 million tonnes of carbon per year. Recognising the impact of deposit age on the net accumulation rate, and the fact that some peatlands may be approaching a steady state, more recent accumulation estimates range from 99 to 194 million tonnes of carbon per annum (Immirzi *et al.*, 1992).

Carbon Flux Element	Mt C/yr
Global Flux	213,000
Terrestrial Flux	120,000
Anthropogenic Emissions	7,700
Peatland Emissions	128

Derived from Bolin (1986)

Figure 3: Peatland Emissions and the Global Carbon Flux

Within Europe, peatland areas and their developmental status are more clearly defined, although significant gaps still exist in the information on the carbon balances under specific end uses. Peatlands occupy 5.0 million hectares within the twelve member states of the present European Union.

Historical development has been particulary intense, with only 36% of the peatland area in the EU remaining in the natural state. The peat industry occupies some 3% of the peatland area, with industry carbon emissions totalling 2.8 Mt C per annum. Assuming an average net carbon accumulation rate of 0.6 t C ha^{-1} yr^{-1}, peatland areas in the natural state recapture 1.1 million tonnes of carbon each year.

	EU - 12	EU - 16
Peatland Area (M ha)	5.0	25.1
Uses - Natural State	36%	54%
- Agriculture	44%	15%
- Forestry	17%	30%
- Peat Industry	3%	<1%
Industry Emissions (Mt C/yr)	2.8	4.5
Natural Capture (Mt C/yr)	1.1	5.8

Figure 4: Comparison of Peat Utilisation in the Present and Future European Union

Expansion of the European Union, through the accession of Norway, Sweden, Finland and Austria, will greatly increase the area occupied by peatlands and alter the proportionate use (Fig. 4). The peat industry will occupy less than 1% of the Union's peatland area, with industry carbon emissions reaching 4.5 Mt C per year. Assuming average net accumulation rates of 0.4 t C ha^{-1} yr^{-1} for areas in the boreal zone, and 0.6 t C ha^{-1} yr^{-1} for areas further south, peatlands in the natural state would be capable of sequestering 5.8 million tonnes of carbon each year.

In the Republic of Ireland peatlands occupy 1.18 million hectares, or 17% of the country's land surface. Just under half of the peatlands remain in a natural, or undeveloped, state. An outline of the main peatland types, and their present application, is presented in Figure 5. The peat industry in Ireland, both State-sponsored and private, produces over 3 million tonnes of dry matter each year, which results in the release of 1.7 million tonnes of carbon.

Assuming an average net carbon accumulation rate of 0.6 t C ha^{-1} yr^{-1} for raised bogs and Atlantic blanket bogs in a natural state, and for a proportion of the cutover blanket bogs under active recolonisation, and 0.4 t C ha^{-1} yr^{-1} for montane bogs, annual capture totals 312,000 tonnes of carbon. This annual capture is equivalent to approximately 18% of the carbon released by the peat industry.

PEATLAND TYPE (000 ha)		PEATLAND USE (000 ha)	
Raised Bog	311	Natural State	565
Atlantic Blanket	337	Agriculture	120
Montane Blanket	437	Forestry	180
Fen	92	Peat Industry	93
		Cutover/Marginal	219
	1,177		1,177

Figure 5: Principal Peatland Types and Uses in Ireland

With under half of its peatland area remaining in a natural state, and peat industry carbon emissions considerably exceeding the natural sink capacity, Ireland makes a comparatively intensive use of its peatland resources. This contrasts with other European peat producers, such as Finland and Sweden, which have positive national peatland carbon balances, owing to their large undeveloped reserves. The range of policies that have supported this intensive peatland development in Ireland, that are presently in operation, and that may be applicable in an era of changing environmental objectives, are examined in the final section.

5. Peatland Policy Issues

A. Historical Policies

Hand-won sod peat, or turf, has been used as a fuel in Ireland since at least the Roman period. Wood was abundant up until medieval times, but with the destruction of the native forests, peat became the only fuel readily available to the Irish poor. Considerable early interest was also shown in the reclamation of peatlands for agriculture. In 1716 the British Government passed an act (2. Geo. 1. 12) 'to encourage the draining and improving of unprofitable bogs and low-grounds' (Cooke, 1970). This act, together with successive studies and reports - the Bog Commission (1809 - 1814), the Irish Peat Enquiry Committee (1918), and the Special Committee on Peat (1921) - proved ineffective, as sufficient funds were never provided to support their recommendations.

In 1933 a separate section was created, within the Department of Industry and Commerce, to organise the production and sale of turf. This was followed in 1934 by the establishment of the Turf Development Board, a limited company financed by grant-in-aid from the Exchequer. During the Second World War coal imports into Ireland fell dramatically and a fuel emergency campaign, under the direction of a Turf Executive Committee, was instigated. The campaign dramatically increased the amount of hand-won sod turf produced annually.

Towards the end of the war a comprehensive plan, for the development of the country's peat resources using mechanical methods, was drawn up by the Turf Development Board. The plan was submitted to the government in December 1944, approved in principle, and published a year later in the form of a White Paper. It received formal legislative sanction in the Turf Development Act, 1946, which provided for the dissolution of the Turf Development Board and the establishment of Bord na Móna, a statutory corporation mandated with the development of Ireland's peat resources.

In the post-war era, energy policy in Ireland was based around two central imperatives: the optimal development and use of indigenous energy resources; and security of energy supply. Successive administrations supported the expanding role of peat, particularly its use for power generation. During the early 1970s, following the initial steep increases in international oil prices, peatland areas previously considered uneconomic, owing to their smaller area or their peripherality, were reappraised and brought on stream. The Turf Development Act, 1981 provided grant assistance for private bog development. Under this scheme 1,478 groups, companies and individuals were approved for grant aid, ultimately resulting in a capacity to produce 1.3 million tonnes of sod turf annually for fuel purposes. During the same period, grants were also made available to householders for the installation of solid fuel heating appliances.

B. Current Policies

The prevailing Irish energy policy has three primary objectives (Irish Government, 1993a):

(i) the supply of a choice of fuels as efficiently as possible;

(ii) the consumption of this energy in an efficient manner; and

(iii) the production of as much of the national energy requirement from indigenous sources as is economically possible.

The present administration, in its 'Programme for a Partnership Government', January 1993, indicated its continuing support for the use of peat in the energy sector. This was reinforced in the policy document 'Employment Through Enterprise', published in May 1993, which stated that

> The availability of peat, as possibly the only significant indigenous fuel after the turn of the century, gives this sector an essential strategic dimension in our national energy security policy (page 17).

The higher carbon emissions per unit of useful energy, from peat-based fuels, are fully recognised in the National CO_2 Abatement Strategy (Department of Environment, 1993). This strategy reaffirms the importance of peat fuel for the security and flexibility of energy supply; recognises that the peat industry is a major employer in midland and western areas; and indicates that, while there will be a certain natural decline in peat consumption,

> measures could not be contemplated which would abruptly compromise the role of indigenous peat in Irish energy supply (page 3).

C. Policies For the Future

I believe that future policies for the management of Irish peatlands should be based on a rational use of our natural resources, with specific areas identified for each of the sectoral interests - peatland conservation and scientific and ecological research, fuel and non-fuel peat production, agriculture, afforestation, and wildlife, landscape and tourism interests. Within the peat industry, the positive attributes associated with security of energy supply, employment and socio-economic benefits must be balanced against increasing environmental awareness, public concerns regarding the potential impacts of climatic change, and the likelihood of changing political, strategic and environmental objectives.

Recognising that the release of carbon is an inherent element of peat industry activities, and that in Ireland the annual release exceeds the sequestration capacity of the intact peatlands, industry-related policies for the future must examine the scope for improving the overall peatland carbon balance:

* In the power sector, carbon release per kilowatt hour generated would be reduced through the blending of peats to provide a uniform feedstock fuel that supports efficient utility operation; through the retirement of the older, lower efficiency peat-fired units when they reach the end of their design life; and through the application of state-of-the-art technology in any replacement capacity, such as the 120 MW fluidised bed combustion unit recently proposed for the East Midlands region.
* On the solid fuels side, the use of enclosed appliances would greatly improve efficiency, and reduce carbon release, when compared with open fires. Such a development would be fully in accord with the governmental strategy of reducing the net energy requirements associated with space heating (Department of Environment, 1993).
* In the horticultural industry, the increased availability of recycled organic

materials, promoted under a number of European directives aimed at curtailing waste and increasing the level of materials sustainability, would reduce primary peat requirements. While recycled or composted materials do not have the same quality characteristics or exchange capacities as horticultural peats, their incremental addition to non-critical growing media and soil mixes would help to reduce overall carbon release.

* The restoration of industrial peatland areas to effective carbon sinks will help to redress the peatland carbon balance. Net carbon gains may be expected over the initial 50 to 200 years for areas planted with coniferous species, through careful management of the water table. However, greater long-term carbon accretion would be achieved through the restoration of cutover areas to an earlier (fen) stage in the hydrosere, when net carbon accumulation rates in mire ecosystems are at their highest (Fig. 2).

Finally, what are the likely global impacts of the projected climatic changes on the peatlands themselves?

Historically, peatlands have responded to previous climatic changes, and any future temperature increases may be expected to affect the balance of photosynthesis and decomposition, particularly in the middle to high latitude mires. Immirzi *et al.* (1992) consider that climatic warming could both renew peat accumulation in regions covered by continuous permafrost, and expand peat formation into new landscapes further north. A drop in water tables at middle latitudes, however, under a warmer and drier climate, would lead to greater acrotelm depth and a loss of net carbon sequestering capacity. The northwards migration of specific mire types is supported by Tolonen *et al.* (1992), but climatic change models are not yet sophisticated enough to allow assessment of whether greater carbon accumulation in the north would be more than offset by increased decay further south.

References

Armentano, T.V. and Menges, E.S., 1986. Patterns of Change in the Carbon Balance of Organic Soil-Wetlands of the Temperate Zone, *Journal of Ecology*, Vol. 74, 755 - 774.

Bolin, B., 1986. How Much CO_2 Will Remain in the Atmosphere? In Bolin, B., Döös, B.R., Jäger, J., and Warrick, R.A., (Eds.), *The Greenhouse Effect, Climatic Change and Ecosystems*, 93-155, SCOPE 29, Chichester, J. Wiley & Sons.

Cannell, M.G.R., Dewar, R.C. and Pyatt, D.G., 1993. Conifer Plantations on Drained Peatlands in Britain : A Net Gain or Loss of Carbon? *Forestry* Vol. 66, **4**, 353-369.

Cooke, J., 1970. Bog Reclamation and Peat Development in County Kildare, *Journal of Co. Kildare Archaeological Society*, Vol. 14, **5**, 576 - 625.

Couch, G.R., 1993. *Fuel Peat - World Resources And Utilisation*, London, IEA Coal Research.

Department of the Environment, 1993. *Ireland Climate Change - CO_2 Abatement Strategy*, Dublin, Government Stationery Office.

Immirzi, C.P. and Maltby, E. with Clymo, R.S., 1992. *The Global Status of Peatlands and Their Role in Carbon Cycling*, London, Friends of the Earth.

Irish Government, 1993a. *National Development Plan 1994 - 1999*, Dublin, Government Stationery Office.

Irish Government, 1993b. *Employment Through Enterprise*, Dublin, Government Stationery Office.

Martikainen, P.J., Nykanen, H., Alm, J., and Silvola, J., 1993. Changes in Vegetation and Fluxes of Carbon Dioxide, Methane and Nitrous Oxide Due to Forest Drainage of Mire Sites of Different Trophy, *(In press)*.

Tolonen, K., Vasander, H., Damman, A.W.H. and Clymo, R.S., 1992. Rate of Apparent and True Carbon Accumulation in Boreal Peatlands, *Proceedings 9th Int. Peat Congress*, Vol. 1, 319-333, Uppsala, International Peat Society.

CLIMATE CHANGE AND COASTAL MANAGEMENT

Robert J. N. Devoy
The Coastal Resources Policy Centre (GEO-CRC)
Department of Geography
University College Cork

The position and extent of the coastal zone upon continental margins is, in the first instance, defined and regulated by the volume and levels of water in the ocean basins. In turn this *sea level* (together with ocean water volume), and its variations through time, are intimately linked to the dynamics of the ocean - atmosphere system: the system working as a forcing agent in sea-level changes. This relationship has been recognised for a long time (eg. Barth and Titus, 1984; Devoy, 1987) and climate change is seen as a primary cause of changes in sea level on a wide range of scales: in time, from hours to 10^3 years and in height amplitude, from mm to 10^2m respectively.

This understanding by coastal scientists of the fundamental role of climate in controlling changes in the coastal zone, led them to be amongst the first to define the possible environmental implications of future greenhouse warming. An extensive body of research literature, practical management studies of the coastal consequences of climate change, and accompanying contingency planning documentation, has resulted (see Devoy, 1992; Peerbolte *et al.*, 1991). This is reflected in the first Intergovernmental Panel on Climate Change (IPCC) documents dealing with the subject (Misdorp *et al.*, 1990) and also in subsequent updates. These together devote a substantial amount of attention to the spatial repercussions of climate change on the coastal zone, underlining the significance of climate forcing on coastal change (i.e. via inundation and storms) as an environmental management issue. The reason: approximately 40% of the world's population lives in close proximity to present sea level and c.66% of these people live within ~80km of the coastline. Although the coastal zone is by definition a dynamic environment, and climate is only one of a range of linked process controls influencing coastal functioning, the outcome of change at the coast is unerringly the same: in some way *people* are affected, hence the interest in coastal zone management.

In 1978 Mercer the attention of scientists and public alike to the possible links between future global warming, ice melt and sea-level rise. Estimates from Mercer's work of a 5m plus global sea-level rise into the next century with climate warming has long since been abandoned, as the nature of ice - ocean - atmosphere interactions have become better understood. Current IPCC and linked estimates of sea-level rise with standard CO_2 - led climate warming scenarios now give values of <0.5m rise by 2050 (Warrick et al., 1993). Linked ocean modelling work also shows the likelihood of distinct regional variations in these values for both the central ocean and ocean margin (continental shelf) environments (Devoy, 1990). Further, increasingly a stress is being laid in such studies upon the error margins accompanying the estimates, given the uncertainties which still exist about ice melt rates and the regional patterns of ocean - atmosphere interactions.

This said, a recognition remains that any future global warming could have significant repercussions for coastal functioning and associated changes in coastal position. Indeed the world's coasts have problems of change now (of both erosion and 'unwanted' sediment accretion), which are either led by or exacerbated by climate functioning. For the maritime countries of Europe their coastal environments have attracted increasingly larger numbers of people and economic investment since at least the 1900s. Every year the effects of storms alone on these coasts provide a source of expenditure and constant worry to coastal administrators. In this context the possible significance for Ireland's >6,500km of coast of any future climate change can scarcely be exaggerated.

Ireland's reponse to these issues at the planning and government levels has been an ambivalent one. Perhaps, uncharitably, this may be interpreted as wishing to be seen to be doing something, but in effect doing little. On the positive side, and following the leads taken by coastal and other environmental scientists in Ireland, the Department of the Environment commissioned an appraisal of the subject (Carter, 1990). This report was subsequently incorporated into a broader document upon the implications of climate change for Ireland (McWilliams, 1991). The 1990 report drew upon the work of coastal scientists in Ireland, usefully placing the needs and options for Ireland's coastal environments under any future sea-level rises in the context of European and other groups' thinking (Devoy and O'Mahony, 1990; Carter and Devoy, 1987). The report stated the likely repercussions of climate change for coastal environments, as then understood, together with the planning-policy issues at stake, and the future needs for action in the management and monitoring of coastal areas. Much of the report was in the van of and may be seen to be consistent with international views on approaches to coastal management, as reflected in the contemporary IPCC documents and elsewhere.

At another level in Ireland, Local Government Authorities and related groups have taken action (see IEI, 1991; IPI, 1993). Galvanized by the effects of storm damage in the late 1980s and subsequently, together with community pressure and the associated rise in public discussion of climate change issues, some Local Authorities have initiated county - regional scale coastal studies and localized responses to coastal management issues. This activity has in some instances been undertaken in conjunction with national bodies, at research level (e.g. Eolas/Forbairt and the EC's ECOPRO programme). Independently again, the seminal work of other sectors in coastal research in Ireland (i.e. in Third Level institutions) has continued at a high level, as instanced in the major contributions of UCC (GEO-CRC) to the EC's *Impacts of Climate Hazards on European Coasts,* under the Environment Framework programmes.

In spite of the encouraging stimulus of the DOE report and the positive contributions of the other sectors of the community, this effectively tripartite approach to coastal issues and climate change, typifies the current situation. Sectoral interests, disunity and non-communication appear to rule in developing coastal management responses to the present and likely future needs of coastal environments under changing climate. The role and action of government is muted and appears to be one of serendipity. Little communication or co-ordination of activities that are ongoing exists between central government and local authorities, or particularly with the work of the different research interests. Although government re-organiztion of departmental portfolios may be promising as a platform for tackling these problems (eg. the Marine Institute, the Department of the Marine), they have yet to bear real fruit for coastal management issues.

A more detailed discussion of Climate Change, Sea-level Rise and Coastal Management in Ireland, is given in Devoy (1992).

References

Barth, M. C. and Titus, J. G., 1984. *Greenhouse Effect and Sea-level Rise: A Challenge for This Generation,* New York, Van Nostrand Reinhold.

Carter, R. W. G., 1990. *The Impact on Ireland of Changes in Mean Sea Level,* Programme of Expert Studies on Climate Change, No. 2. Dublin, Department of the Environment.

Carter, R. W. G. and Devoy, R. J. N., 1987. The Hydrodynamic and Sedimentary Consequences of Sea-Level Change, *Progress in Oceanography,* **18**, London, Pergamon Press.

Devoy, R. J. N. 1987. *Sea Surface Studies: A Global View,* London, Croom Helm - Chapman Hall.

Devoy, R. J. N., 1990. Sea-level changes and Ireland, *Technology Ireland,* **22**, 24-30.

Devoy, R. J. N., 1992. Questions of coastal protection and the human response to sea-level rise in Ireland and Britain, *Irish Geography,* **25**, 1-22.

Devoy, R. J. N., and O'Mahony, A., 1990, *Coastal Engineering and Management: A Workshop,* Resource and Environmental Management Unit, University College Cork, 98pp. (Available from the Coastal Resources Centre, UCC).

Institution of Engineers of Ireland, 1991. *Engineering for Coastal Protection,* Dublin, Institution of Engineers of Ireland, 149pp.

Irish Planning Institute, 1993. (IPI): *Planning for Coastal Areas and Inland Waterways: Conference Proceedings,* Dublin, Irish Planning Insititute.

McWiliams, B. E. (Ed.), 1991. *Climate Change : Studies on the Implications for Ireland,* Department of the Environment, Dublin.

Mercer, J. A., 1978. West Antarctic ice sheet and CO_2 greenhouse effect: a threat of disaster, *Nature,* **271**, 321-5.

Misdorp, R., Dronkers, J., and Spradley, J. R., 1990. *Strategies for Adaption to Sea-level Rise,* Report of the Coastal Zone Management Subgroup, Intergovernmental Panel on Climate Change, Response Strategies Working Group, Rijkswaterstaat, The Hague Ministry of Transport and Public Works.

Peerbolte, E. B., de Ronde, J. G., de Vrees, L. P. M., Mann, M., and Barse, G., 1991. *Impact of Sea-level Rise on Society: A Case Study for the Netherlands,* Delft Hydraulics.

Warrick, R. A., Barrow, E. M., and Wigley, T. M. L., 1993. *Climate and Sea-level Change: Observations, Projections and Implications,* CUP.

THE INSURANCE INDUSTRY'S VIEW OF CLIMATE VARIATIONS

A. F. Dlugolecki
General Accident (UK)
J. Fraser
General Accident (Ireland)

Introduction

At the outset it may be useful to outline the qualifications of the two authors of this paper. Dr Dlugolecki is a co-author of the UK study of the potential impacts of climate change, in particular the section on financial services, and is also involved in the IPCC study due for publication in 1995. Mr Fraser has worked for many years in the Irish Insurance Industry, and is currently Chairman of the Property Committee of the Irish Insurance Federation.

The insurance sector comprises many different activities: marketing, underwriting (ie transferring risks for a fee or premium), risk management, settlement of claims, and investment - arising from the financial protection of a wide variety of human concerns: death, health, property, motor and liability to others. Naturally many organisations and specialisms exist to serve these needs, but in this paper we will concentrate on property insurance from the underwriting company's perspective.

Insurance works by collecting together funds from many individuals to pay for the few random losses which could ruin a small proportion of the contributors. The basis for charging the premium is usually historic experience, modified by competitive pressures and likely future changes in the business environment. Any major unforeseen worsening in the frequency or cost of incidents will therefore weaken the insurer's capital base and certainly lead to a rapid rise in premiums and/or a restriction in the cover, or contract terms. The immediate impact may be softened by 'laying-off' some of the cost to reinsurers, but they will also wish to make a profit, and are well-placed to make a judgement on global trends, contrary even to better experience in individual countries, like Ireland.

Recent History of Property Insurance and Weather

The most obvious way in which climate change would affect property insurers is through a change in weather patterns - extreme events such as storms, floods, frost, hail, wildfire or drought usually damage property!

Unfortunately, insurers have not been natural publishers of loss statistics. Very few countries have as much as 30 years data, and Ireland has not compiled data on weather losses, because it has been relatively quiet in recent years. This is most unrepresentative of what has happened in other countries.

Worldwide figures compiled by reinsurance companies show that the insurance cost of major windstorms rose from $500 million per year in the 1960s, to $1700 million in the 1980s. (Dlugolecki, 1992). In the 1990s it accelerated, reaching over $10 billion per year. The cost of Hurricane Andrew alone in 1992 was about $16 billion (Leggett, 1993).

The UK has seen a similar pattern, culminating in the storms and drought of 1990. (Dlugolecki, 1991). Since then there have been a number of serious local floods, and a near miss, when the worst storm recorded in the North Atlantic destroyed the oil tanker Braer on the Shetlands in January 1993.

These huge shifts in loss experience can be partly attributed to non-weather effects: population settlement in exposed areas, increasing wealth, more delicate property (e.g. satellite dishes, Hi-Fi sets), wider insurance contracts, and unfortunately, fraudulent claims. However, a significant contribution is due to the weather, as the patterns of rainfall and windstorm have altered. (Dlugolecki, 1992).

Insurance Opinions on Climate Change in General

Some reinsurers have expressed grave concern at the prospect of Global Warming (Leggett, 1993). However, it is true to say that weather patterns are still within historical variability, and in a territory like Ireland, where there have been no major 'events', there is still some scepticism. This is understandable when there is still substantial uncertainty about how key features of the global climate work: cloud cover, natural eco-feedback, and oceanic currents for example. Yet even the cynics are shifting their position in the face of occurences like the Mississippi floods, and the Sydney bushfires which show the sort of thing which might face us with climate change in future.

The climate modellers have still not been able to produce consistent rainfall scenarios, and detailed tabulations of the parameters which vitally concern insurers are still not available: windstorm patterns, daily temperature, and daily precipitation (Dlugolecki, 1991). Without this it is difficult to plan in detail.

It is all very well to talk of planning for a future date like 2010 or 2040 (Maunder, this volume), but in fact the climate will continue to drift until draconian measures are taken to control CO_2 emissions (Leggett, 1993).

One thing which *is* apparent is that the use of 'return periods' whereby historic data on storms or floods are used to calculate the frequency of various levels of severity, is a dangerous practice. Generally the available observations on river flow or other environmental variables are fairly recent, and even historical variability is not well-reflected in a time span of decades.

Climate Change in Ireland

With its exposure to the Atlantic, windstorm is an obvious hazard in Ireland, but most of the expensive property is in the east or is in sheltered areas, as in Cork, and this is not seen as a new threat. However, the global patterns show that the new events could exceed historical costs many-fold, and the main danger here is complacency.

Flood is also a worrying feature, when taken in conjunction with rising sea-level and storm-surge, reclamation projects, faster runoff due to urbanisation, and potential changes in rainfall patterns (McWilliams, 1991). Cork and Belfast have suffered local flooding over 200 times each in historical records, but the major exposure is in Dublin and Portmarnock. In fact the Baldoyle floods of June 1993 gave a timely indication of this, costing £IR30 million. Similar events have been witnessed in the UK, both in summer and in winter in 1993 and 1994.

The prospect of lower summer rainfall could give rise to subsidence claims, as has happened in the UK several times since 1976. It is true that the likelihood of wetter winters may offset this, but if the winter rains were to fail, then the circumstances could damage domestic dwellings significantly.

Immediate Strategies

Because the UK was hit repeatedly by extreme weather, insurers and reinsurers there have reacted quickly. In the UK, domestic building premium rates have increased by 100% in ten years. At the same time insurers have begun to differentiate between neighbourhoods purely because of weather factors, and more recently some insurers are considering the construction features of individual domestic buildings. It should be noted that due to a 1961 agreement between insurers and government, almost all households can obtain flood insurance at reasonable rates.

Ireland is able to learn from the UK experience, and insurers here are considering a number of actions which collectively will prepare the way for meeting the inevitable problems which will arise:

A) Through the Loss Prevention Council, based in England, to re-assess construction standards regarding storm resistance and flood protection. Admittedly this will not affect existing housing at first, but repairs could be performed to the new higher standards also.

B) Lobby for tax exemption on loss equalisation reserves, to allow insurers to offset good years against bad, instead of funds passing out of the country to reinsurers. The UK Government recently took the initiative on this very subject, to assist in providing a level playing field with continental competitors (Department of Trade and Industry, 1993).

C) Raise the level of storm excesses to some extent in order to reduce the huge volume of claims in a superstorm, without unduly straining the policyholders' purse.

D) Initiate consultation with the Department of the Environment regarding storm drainage and coastal defences. Proper physical preparation would mitigate damage.

E) Maintain a dialogue with reinsurers to ensure that Ireland does not get lumped in with other less fortunate but larger territories like the UK for the purpose of reinsurance charges.

F) Monitor scientific progress to ensure that as good a quality of information as possible is applied to Ireland specifically.

G) Collate statistics on weather losses for public information, and to assess the impact of specified severities of event.

H) Publicise the benefits of insurance cover, and the relatively good value to the consumer in Ireland, where property premium rates are much lower that the UK. The wider the insured base, the bigger the fund to meet losses, and the less need for disaster relief handouts after a catastrophe.

Implications of the Absence of Insurance

When insurers cannot achieve a balance between premium versus claims and expenses, or when the potential loss is so unpredictable that it could cause the insurer's ruin, then the only rational course is not to underwrite the risk. This can mean significant difficulties for other parties:

A) Government will have to step into the breach with disaster relief, diverting public resources from their intended use.

B) Homeowners will not receive full compensation for their losses, and will have to manage their domestic crisis unaided by the expertise of insurance organisations.

C) Financial institutions may find that their loans are no longer good security, since physical damage cannot readily be restored without an insurance cover to generate funds for repair.

For those reasons it is vital that the dialogue between insurers and other parties, particularly Government, commences and is a healthy one.

The Insurance Contribution to CO_2 Abatement

A variety of measures in the management of energy use, transportation, waste disposal and afforestation have been outlined in a recent government paper (Department of Environment, 1993). At first sight the insurance industry can not make major contributions, since it is a service process, not primary or manufacturing.

The most effective ways may be as follows:

A) **Transportation**: in their rating of motor risks, insurers currently already penalise those factors which will generate a heavy use of energy through high mileage, urban movement, and rapid acceleration. Thus a hidden benefit of this is encouraging energy conservation.

B) **Paper**: this is one of main by-products of insurance! It is also energy-intensive; therefore it seems the modern trend to large-print, colourful and glossy documents is a backward step in terms of conservation. This may be offset in future by increased use of the telephone, electronic mail and recycled paper. No insurer has yet marketed itself as 'green'.

C) **Investment Strategy**: an insurance company has a duty to its share-holders and policyholders to invest their funds for a good and secure return. There may soon be opportunities and threats for investment arising from Global Warming. For example, forestry, mass transport and solar batteries may be positive, and oil, private vehicles, and coastal sites may be negative. However, the normal principles of investment appraisal must still apply to ensure that an economic return is achieved.

D) **Lobbying**: finally, if insurers can be convinced that the stability of their activities is threatened by climate change, they may become so bold as to press governments for a faster response to global warming. (Leggett, 1993).

Conclusion

In this brief foray, we have shown that the Irish insurance industry has been fortunate as regards weather impacts in recent years. To its credit, it has noted events abroad, and the scientific consensus on global warming, and is now actively considering a programme of immediate actions. This will hopefully avert a crisis at a later stage, if it is integrated with initiatives, especially from Government. The strategies could be even more strongly focused, given more accurate and detailed information from climate models.

Insurers already contribute to CO_2 abatement in their motor rating approach, but, there is a variety of other options which they can consider, including reduced use of paper, investment strategy, and lobbying.

References

Department of the Environment, 1993. Ireland: Climate Change: CO_2 Abatement Strategy, Dublin, Stationery Office.

Department of Trade and Industry, 1993. Trade Equalisation Reserves for Insurance Companies Non-Life Business - A Consultation Document, London, HMSO.

Dlugolecki, A., 1991. Financial Sector, In *The Potential Effects of Climate Change in the United Kingdom*, London, HMSO.

Dlugolecki, A., 1992. Changing Weather Patterns, Insurance Institute of Dublin (unpublished).

Leggett, J., 1993. *Climate Change and the Insurance Industry*, London, Greenpeace.

McWilliams, B., 1991. Executive Summary, In *Climate Change: Studies on the Implication for Ireland*, Dublin, Department of the Environment.

CLIMATE SCENARIOS FOR IRELAND

John C. Sweeney
Department of Geography
St. Patrick's College, Maynooth

General Circulation Models and Future Projections of Irish Climate

Global scale general circulation models (GCM) can currently only provide a rough guide to the likely future course of Irish climate in an enhanced greenhouse effect-warmed world. Such computer models use basic physical laws to describe changes in momentum, heat and water vapour as a consequence of atmospheric motion and have evolved from elementary and rather crude constructs to complex and demanding affairs which require some of the most powerful computing platforms in existence. For the purposes of simulating the effects of a hypothetical change in climatic inputs, the atmosphere is divided into a number of vertical levels (up to 19) and spatially organised into a series of horizontal grid points (300-1,000km apart). For each of these layers, and each grid point, a number of climatic parameters are calculated by running the model forwards in a series of simulated time steps (of the order of 30 minutes) starting from some initial condition. Normally equilibrium results are obtained after a few tens of years and comparisons between the equilibrium climate and a control run can be made.

The technique has only become possible due to the advances which have been made in computing speed over the past two decades. Typically, in the late 1970s the most powerful computers could cope with approximately 10 million floating point operations per second (MFLOPS) while today the figure is approximately 10,000 MFLOPS. This is important since each iteration of the model requires the retrieval, calculation and storage of approximately 10^5 numbers for each grid point for each atmospheric variable. Such advances mean that seasonal and regional details are just beginning to emerge as output possibilities, as is the vital question of future climatic variability.

Early generations of GCMs showed considerable inconsistencies, particularly for low latitude pressure and high latitude precipitation simulations. These have now begun to diminish. It is also now possible to run models which incorporate transient changes, i.e. annually changing a parameter such as CO_2 instead of arbitrarily doubling it in one initial step. This provides a much more realistic view of how gradual changes in climate are likely to occur.

It is however, fair to say that models have not yet evolved to cope adequately with a variety of feedback effects, and this continues to limit their utility, particularly in providing the regional scale forecasts which policy makers require. They also have been less than satisfactory in incorporating the interaction of the ocean with the atmosphere and in coupling the two systems' role in regulating climate change.

Feedbacks

A positive feedback acts to magnify an initial perturbation while a negative feedback dampens change. The most important feedbacks implicated in climate change are those relating to ice-albedo, water vapour and clouds.

Ice-albedo

The current planetary albedo (or reflectance fraction) of 0.30 is determined largely by the amount of snow and ice existing. Such surfaces may reflect up to 90% (albedo 0.90) of incoming insolation and as a consequence play a large role in determining the net radiation balance of a place. During a period of warming it can be expected that the melting of snow and ice at high latitudes will decrease the albedo in these locations, further enhancing warming and accelerating the retreat of the ice margin. Such a positive feedback mechanism may partly explain the newly-discovered rapid climate warming which occurred at the end of the last glaciation.

Water Vapour

Water vapour is the main greenhouse gas and absorbs terrestrial radiation in the 5-7μm wavelength range. Should temperatures rise as anticipated, increased evaporation from the oceans and other moist surfaces might be expected to reinforce a warming trend, a further example of a positive feedback mechanism.

Clouds

Clouds are the source of greatest uncertainties with GCMs since it is not even clear in which direction feedback operates with them. Highly reflective to incoming insolation, clouds might be expected to behave rather like a snow or ice-covered surface, and so greater evaporation in a warmed world leading to more clouds might be expected to increase the planetary albedo and slow cooling - a negative feedback effect. However, additional water vapour amounts might also be expected to absorb more terrestrial radiation, enhancing warming - a positive feedback consequence. It is also important to realise that changes in cloudiness may not necessarily be the same as changes in cloud cover. More stratiform clouds may mean greater cloud cover than more cumuliform clouds, an important consideration for deciding the direction of the overall feedback effect.

To date GCMs have been conspicuously unsuccessful in handling cloud formation, especially in terms of height. Clouds occur at scales below the resolution of all the current generation of models and cannot be easily incorporated other than as parameterised variables.

Feedbacks in Combination

Since several feedbacks are apparent, some positive, some negative, there is a need to determine their additive effect. If this is achieved it will determine how sensitive the climate system is to CO_2 forcing.

One way of addressing this is to ask the question: 'What temperature change ΔT will occur in response to a prescribed (e.g. doubled CO_2) net radiative flux change, ΔQ, across the tropopause ?' To answer this the following expression can be considered (Henderson-Sellers and McGuffie, 1987):

$$C\,[\,\delta(\Delta T)/\delta t] + \lambda\Delta T = \Delta Q$$

where: C is the system heat capacity,

and: $\lambda\Delta T$ is the net radiative change at the tropopause resulting from the internal characteristics of the climate system; t is time

An appropriate value for λ is λ_B which is the value the earth would have if it was a black body with its present day albedo, so that:

$$\lambda_B = 4\,\sigma T_e^{\,4} = 3.75 \text{ W m}^{-2}\text{ K}^{-1}$$

where: σ is the Stefan-Boltzmann constant,
and: T_e is the earth's effective temperature.

The total sensitivity of the earth's climate can now be obtained by adding the internal feedback factors as follows:

$$\lambda_{Total} = \lambda_B + \lambda_{water\ vapour} + \lambda_{ice\ albedo} + \lambda_{clouds}$$

Some of these quantities have been established. For doubling of CO_2, it is known that $\Delta Q = 4.0$ W m^{-2}, while $\lambda_{water\ vapour}$ = -1.7 W $m^{-2}K^{-1}$ and $\lambda_{ice\ albedo}$ = -0.6 W $m^{-2}K^{1}$.

Thus:

$$\lambda_{Total} = 1.45 + \lambda_{clouds}$$

Current estimates for λ_{clouds} range from 0 to -0.8 W $m^{-2}K^{-1}$ (Henderson-Sellers and McGuffie, 1987).

Substituting λ_{Total} into the first equation above, the equilibrium temperature $[\delta(\Delta T)/\delta t = 0]$ for doubled CO_2 can be calculated. A zero value for λ_{clouds} gives ΔT=2.8°C, while a value of -0.8 W $m^{-2}K^{-1}$ gives a temperature change of 6.1°C. This range of results clearly implies that cloud feedbacks must be better understood before GCM results may be accepted with confidence.

Response times

Some components of climate adjust themselves very quickly in response to a given perturbation. For the lower atmosphere, equilibration may occur in a few days, while for the deep oceans and ice sheets several centuries may be required. This poses problems for the modeller since the oceans act both to remove CO_2 from the atmosphere and also to absorb and redistribute heat throughout their great volume. Indeed the upper three metres of the ocean store as much heat as the entire atmosphere, making the oceans an essential part of any climate model. Early GCMs treated the oceans simplistically and it is only recently that coupled ocean-atmosphere models have had any success in meshing these two vastly different systems.

Model Estimates for Irish Climate

As models have improved, their estimates of total global temperature change consequent on effective CO_2 doubling have reduced and greater consistency between the major products has become apparent (Figure 1). This is partly a consequence of the models' ability to handle new negative feedback influences, such as the effects of stratospheric ozone depletion and of greater plant growth and carbon storage. These have been incorporated into the latest generation as has the role of other anthropogenic influences such as increased sulphate aerosols from the industrialising countries of the Third World. The net effect is to diminish the magnitude of warming suggested by earlier generations. Thus, while global warming of 4°C by mid century was widely suggested in the late 1980s (Hansen *et al.*, 1988), the 1992 supplementary IPCC report suggests a best estimate for the period 1990-2100 of 1.5-3.5°C (Leggett *et al.*, 1992).

For areas in the vicinity of Ireland, there is broad agreement among the major models that a similar range of figures to the global values may be appropriate. However, the coarseness of the typical GCM grid size does not provide adequate resolution for policy makers. An approach which has helped rectify this deficiency has been to use the output from a particular set of grid cells to drive a second spatially-detailed regional model. This nested regional model approach is still in its infancy but it does appear to offer promise in providing regionally detailed output (Figure 2).

It can be suggested that the climate of Ireland may be relatively slow to respond to greenhouse forcing due to the moderating influence of the Atlantic Ocean. The Atlantic is a major heat sink, with summer heat being dissipated each winter through a depth in excess of 200m, and over a large area off the west coast of Ireland this value exceeds 500m (Rowntree, 1990). Thus, even if the European continent warms up as it did in the Holocene post glacial optimum, the existence of a cool sea surface around Ireland will inhibit warming in summer, just as it did in the post-glacial optimum climates of 6000 B.P.. During winter, since isotherms run north-south across Ireland, a

Figure 1: Surface temperature change consequent on doubling effective CO_2 as shown by four coupled ocean-atmosphere transient general circulation models

(a) Geophysical Fluid Dynamics Laboratory, Princeton

(b) Max-Planck Institute, Hamburg

(c) National Centre for Atmospheric Research, Boulder

(d) United Kingdom Meteorological Office, Bracknell

(after Gates *et al.*, 1992)

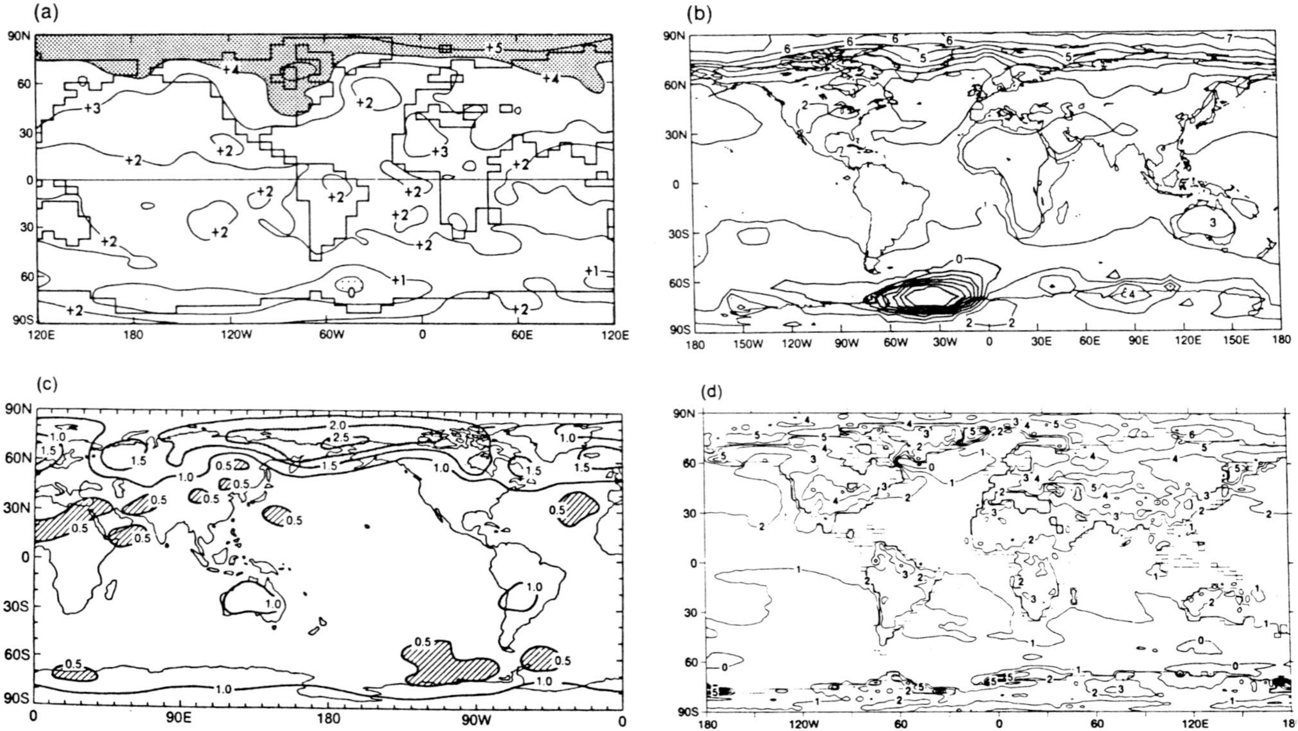

Figure 2: Increase in January temperatures in Europe as a result of effective doubling of CO_2

(a) National Centre for Atmospheric Research General Circulation Model

(b) Nested Mesoscale Model (after Gates *et al.*, 1992)

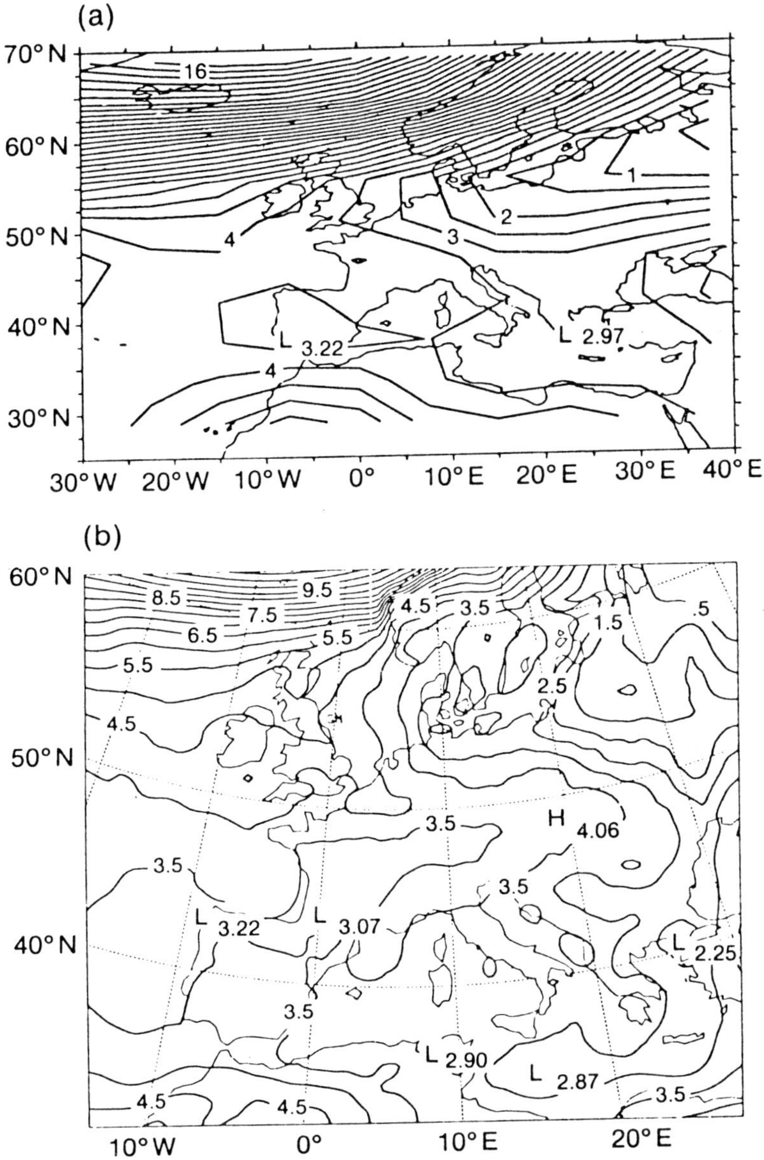

poleward shift in winter isotherms will have less effect than further east in Europe. Some support for these suggestions can be found in the latest generation of transient GCMs which incorporate deep ocean mixing (Stouffer *et al.*, 1989; Washington and Meehl, 1989). They are also supported by the work of Karoly (1987) who in a study of actual temperatures at the 700mb level between 1960-80, showed that warming was apparent almost everywhere except the north Atlantic and Pacific Oceans (where considerable winter mixing depths also exist). If indeed the Atlantic around Ireland is slow to warm, the consequences for Irish climate will be determined primarily by changes in circulation frequencies.

Predictions of changes in precipitation are much less consistent than temperature and tend to be founded on the atmosphere's capacity to hold approximately 7% more water vapour per 1°C rise in temperature. Since the high middle latitudes warm most, the highest increases in precipitation are suggested as occurring here, particularly in winter (Figure 3). However, the models are not consistent in identifying the precise areas where precipitation changes may occur and locations such as Ireland tend to show contradictory signals in different models. This is especially the case in summer. This is unfortunate since summer precipitation is of critical importance for Irish water resource management in areas ranging from agriculture to effluent dilution. Indeed, even in the absence of precipitation changes, significant increases in evapotranspiration will occur from additional warmth (Lockwood, 1993). In Ireland such increases, if accompanied by even modest reductions in summer precipitation, will lengthen considerably the periods of soil moisture deficit with potentially very significant impacts on crop yields possible on occasion. Figure 4 shows the effects on moisture balances of a simulated 3°C rise in summer temperature accompanied by a 5% reduction in summer precipitation on typical inland and coastal locations in Ireland. (Evapotranspiration estimates were estimated only for the summer months using the approximation method of Crowe (1971))

Evidently, while some confidence may be possible that a doubled CO_2 climate in Ireland will be accompanied by winter and summer temperatures approximately 2°C warmer, little confidence exists in projected precipitation estimates for Ireland at present. In particular the important issue of summer precipitation changes cannot be addressed satisfactorily by the models as yet, and so an alternative approach is necessary. One such strategy involves an analogue climate approach.

Projecting Future Irish Summer Rainfall using a Synoptic Circulation approach
Short term future climate change in Ireland will be determined principally by changes in circulation frequency. Indeed significant changes in frequency have

Figure 3: Changes in precipitation consequent on doubling of effective CO_2 as modeled by the UK Meteorological Office model

(a) December, January, February (b) June, July, August

(after Gates *et al.*, 1992)

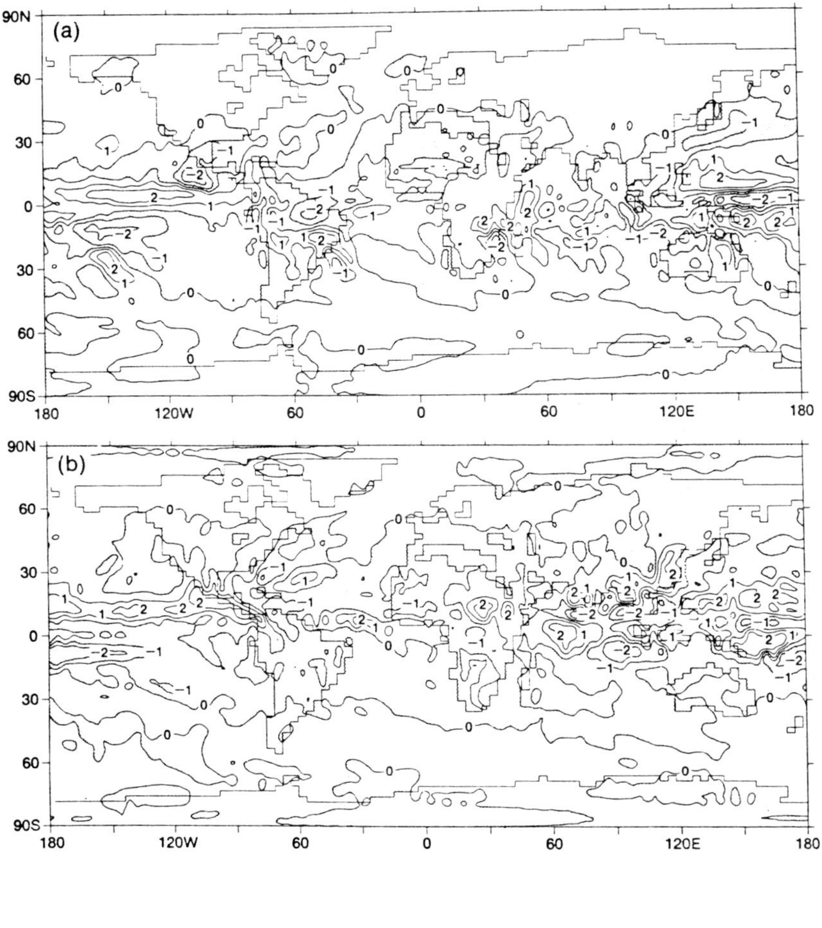

Figure 4: Present and projected soil moisture balances for Birr and Valentia

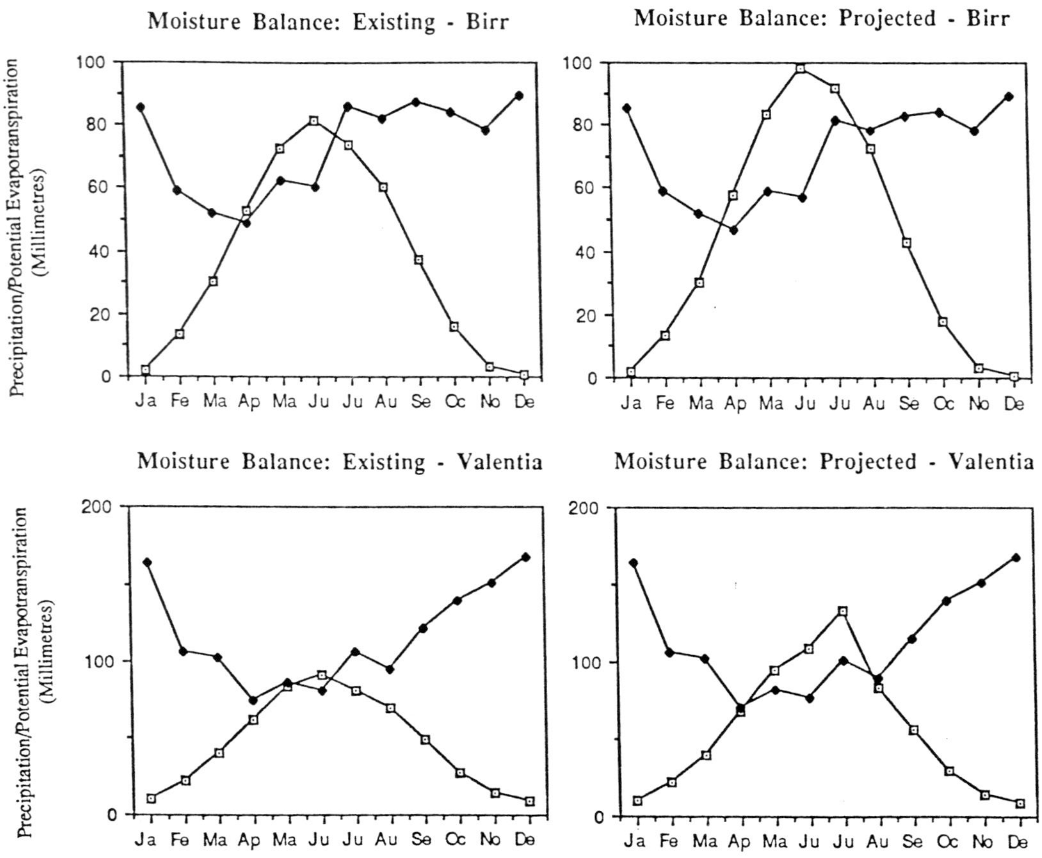

already occurred (Briffa *et al.*, 1990) and have been implicated in changes in rainfall geography within Ireland (Sweeney, 1985). If present relationships between circulation types and rainfall geography can be established, it may thus be possible to simulate precipitation patterns for hypothetical scenarios of circulation which may accompany greenhouse-led climate change.

The first requirement is to achieve some form of categorisation of synoptic circulation type for airflows affecting Ireland. The best known such classification was derived by Lamb (1950, 1972) who identified seven primary circulation types: anticyclonic, cyclonic, north-westerly, westerly, northerly, easterly, and southerly. Subsequently he added a further nineteen secondary categories (together with an unclassified category) incorporating more complex circulation types into a daily catalogue extending from 1861 and still updated by the author daily. The categories are considered to be reasonably representative of an area (50-60°N and 10°W-2°E) which includes Ireland, though problems of intra-regional variation do occur (O'Hare and Sweeney, 1992).

Using daily rainfall data from 34 stations in Ireland for approximately a 40 year period, mean rainfall yields were calculated seasonally and annually for each of the Lamb circulation categories. This revealed significant spatial contrasts in rainfall receipt with different circulation types (Sweeney, 1985). For example an anticyclonic northerly airflow produces on average less than 0.5mm of daily rainfall across Ireland while a cyclonic southerly yields approximately 5mm. Similar spatial contrasts are also apparent across Ireland, however, even for the same circulation type (Figure 5).

Present Circulation-Precipitation Relationships

Anticyclonic

This airflow produces little precipitation anywhere in Ireland, as might be expected with stable, subsiding air. Amounts received are generally below 1.0mm. The south-westerly flow of air on the western flank of anticyclones may allow some fronts to skirt western Ireland and slightly higher amounts are in evidence here.

Cyclonic

More than any other circulation type, cyclonic airflow promotes an even distribution of rainfall across Ireland, with only slight reductions away from coastal locations apparent as the oceanic water vapour supply is presumably reduced. Clearly, should an increase in cyclonic airflow frequency increase consequent on global warming, the characteristic west-east contrast in annual rainfall receipt in Ireland would be expected to diminish.

Westerly

A marked spatial contrast between west and easterly locations is apparent

Figure 5: Precipitation yields in Ireland with principal Lamb synoptic circulation categories

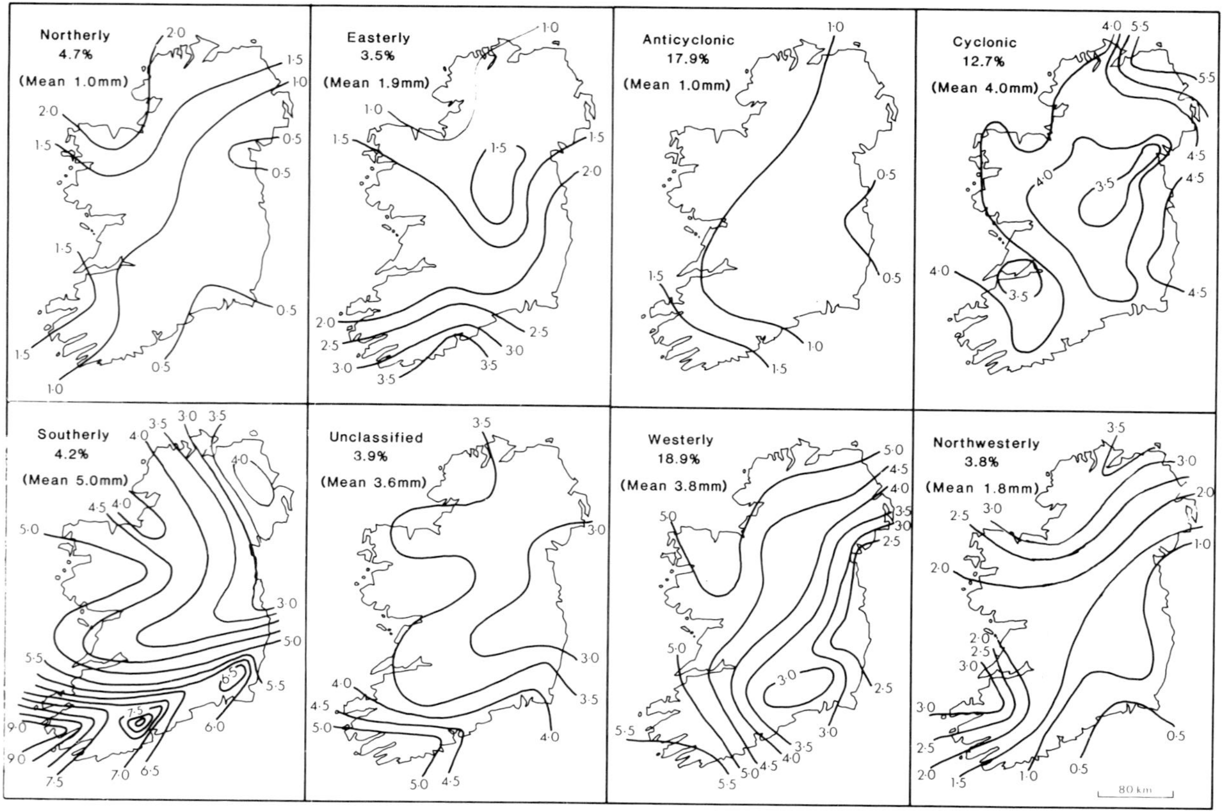

with this airflow which, because of its frequency (18.9%) is therefore responsible for the west-east gradient in the annual rainfall distribution. Parts of Connaught receive three times as much rain with this airflow than sheltered parts of the Leinster coast. Pronounced rain shadow areas are also apparent downwind of the mountain ranges. It can be concluded that the effects of a decline in the westerly circulation, something which has already been occurring for several decades, would again diminish west-east rainfall contrasts. Such a scenario is of course likely as the equator-pole thermal gradient reduces with global warming.

North-westerly

Ridging, behind a depression or by extension of the Azores anticyclone, produces falls of over 3.00mm in the north-west and on exposed coasts in Munster. Elsewhere this is quite a dry airflow, especially along the south coast where it is the driest airflow experienced.

Northerly

With high pressure to the west of Ireland, a northerly airflow occurs at present about 4.7% of the time. Northern coasts in Ulster and Connaught receive most precipitation with significant rain shadow areas to the south of the Mournes and Wicklows.

Easterly

The longest sea passages for easterly air occur along the Celtic Sea. Depressions passing south of Ireland are mostly involved in producing this airflow which may also be accompanied by westward-moving fronts. Co. Cork and much of the south coast get higher yields from easterlies than they do from westerlies. Rain shadows are distinctive of the west side of mountain ranges.

Southerly

Maritime tropical airmasses arrive in Ireland heavily laden with moisture and it is therefore not surprising that southerly circulation types are the wettest airflows to affect Ireland. As air cools passing northwards over the North Atlantic Drift, stratiform cloud is produced and this readily yields copious rainfall when lifting over the mountains of southern Ireland occurs. With any warming of the Atlantic, southerly airflows would become more capable of carrying water vapour, and increases in rainfall especially in southern areas would be likely.

Circulation-Precipitation Relationships for Greenhouse-Effect Summers in Ireland

Changes in circulation frequency, and thus in precipitation, can be expected to occur in Ireland as greenhouse-led warming proceeds. Such changes will dominate the climatic response for the next few decades (Rowntree, 1990). In order to anticipate what impacts are likely on rainfall receipt during summer, one approach would be to examine what mix of circulation types was associated with warm summers in the past. To address this question, the long instrumental record developed by Manley (1974), and since then continuously updated, was utilised. Though symptomatic primarily of temperature conditions in central England, Manley's record is generally held to be fairly representative of general temperature conditions over Britain and Ireland extending back to the mid 17th century, and is the longest instrumentally observed record of temperature available globally. A long record is essential to avoid the distortions which anomalous blocking might produce in a shorter data run.

Summers exhibiting positive temperature departures of +1 (warm) and +2 (hot) standard deviations from the mean were extracted from the Central England Temperature record (Table 1). The average temperature of the latter corresponded to additional summer warmth of 1.7°C, close to the likely warming predicted for Ireland by GCMs within the next forty years. The frequency of circulation types during those years of the Lamb record which met this criteria are shown in Table 2.

Table 1: Warm and hot summers in the Central England Temperature Series

Warm Summers

(>1δ above mean average June/July/August temperatures 1659-1992)

1666,1676,1679

1701,1706,1707,1718,1719,1727,1728,1731,1733,1736,1747,1759,1762,1772,1775,1778,1779,1780,1781,1783,1794,1798

1800,1808,1818,1826,1831,1834,1835,1846,1857,1859,1868,1870,1887,1893,1899

1911,1921,1933,1934,1935,1947,1949,1955,1959,1975,1976,1983,1984,1989

Hot Summers

(>2δ above mean average June/July/August temperatures 1659-1992)

1781,1826,1846,1899,1911,1933,1947,1975,1976,1983

Hot summers in Britain and Ireland are clearly associated with a significant increase in the frequency of anticyclonic circulations and a significant decline in the westerly and cyclonic category. This of course implies blocking and a tendency for the upper westerly circulation to exhibit a more cellular pattern. Such a tendency for increased low zonal index circulations can be expected in

Table 2: Lamb Circulation Frequencies 1880-1992

	All Year	All Summers	Warm Summers	Hot Summers
Unclassified	4.0	3.8	3.3	3.3
Anticyclonic	18.0	19.0	31.1	33.1
AC/N.E	1.3	1.3	1.8	1.7
AC/E	2.5	2.4	3.7	4.3
AC/S.E.	0.9	0.7	1.5	1.4
AC/S	1.1	0.9	1.0	1.4
AC/S/W	0.8	0.8	0.7	0.6
AC/W	4.6	5.4	6.1	4.8
AC/N.W	1.4	2.1	2.8	2.5
AC/N	1.9	2.2	2.7	2.0
North-East	0.9	1.0	0.6	0.3
Easterly	3.5	1.9	2.4	3.0
South-East	1.8	0.7	1.2	1.4
Southerly	4.4	2.4	3.3	3.6
South-West	2.8	2.1	2.1	2.3
Westerly	18.5	17.8	13.3	10.9
North-Westerly	3.8	4.8	3.4	3.7
Northerly	4.8	5.0	2.7	2.0
Cyclonic	13.0	15.9	9.3	9.8
Cy/NE	0.4	0.3	0.2	0.3
Cy/E	1.1	0.9	0.7	0.8
Cy/SE	0.5	0.2	0.2	0.2
Cy/S	1.2	1.0	1.4	1.1
Cy/SW	0.5	0.5	0.3	0.3
Cy/W	3.8	4.4	2.6	3.0
Cy/NW	0.9	1.1	1.0	1.4
Cy/N	1.4	1.3	0.7	0.9

Note : Warm Summers are defined as greater than 1 std. dev above CET mean 1659-1992.

Hot Summers are defined as greater than 2 std. dev above CET mean 1659-1992.

a greenhouse-forced climate as differential warming of the poles more than the equator reduces the equator-pole thermal gradient and diminishes the vigour of the westerly circulation. Declines in the westerly circulation have been occurring over Ireland and Britain for several decades, though short term reversals of this occur from time to time.

The relationship between circulation type and rainfall yield was discussed earlier. This was based on approximately 40 years of rainfall data. During this interval changes in circulation frequency were obviously occurring. However, an examination of decadal breakdowns of the data showed that the average yield did not show significant variations for a particular circulation category over time. Therefore, it is possible to project a hypothetical rainfall climate for each station using the 'hot summer' circulation frequencies. This gives a projection for rainfall receipt by circulation category in a simulated greenhouse-forced climate situation which can be mapped to show regional variations (Figure 6).

Reductions in summer rainfall of in excess of 15% are indicated for Ireland with the largest diminution occurring in the north and east. Belfast, for example, shows a reduction in summer rainfall receipt of approximately 20% using this approach. Along the south coast, by contrast, little change is apparent primarily due to the ability of warm, moist southerly circulations to compensate for the effects of reductions in westerly and cyclonic rainfall contributions.

Increases in summer evapotranspiration of approximately 8% will accompany greenhouse-led temperature changes. Should rainfall reductions occur these are likely to be even more important considerations for water resource management, and Figure 6 thus shows where public water supply, agricultural water management planning, and effluent disposal, concerns might be concentrated.

Some decreases in summer rainfall have in fact occurred in Ireland over the past two decades, though some reversal of this trend occurred in northern and western areas during the 1980s. In Britain, eastern and southern parts also experienced the greatest reductions, and Mayes (1991) has suggested that these are associated with changes in anticyclonic and westerly circulation frequencies. Such changes provide some observational support for the scenario indicated here for Ireland which diverges from the predictions of most GCMs concerning summer precipitation changes.

Conclusion

Greenhouse-led climate change is likely to be detected later in Ireland than in much of the remainder of the northern hemisphere due to the sluggish

Figure 6: Projected summer precipitation in Ireland as % of present as a result of 2°C warming

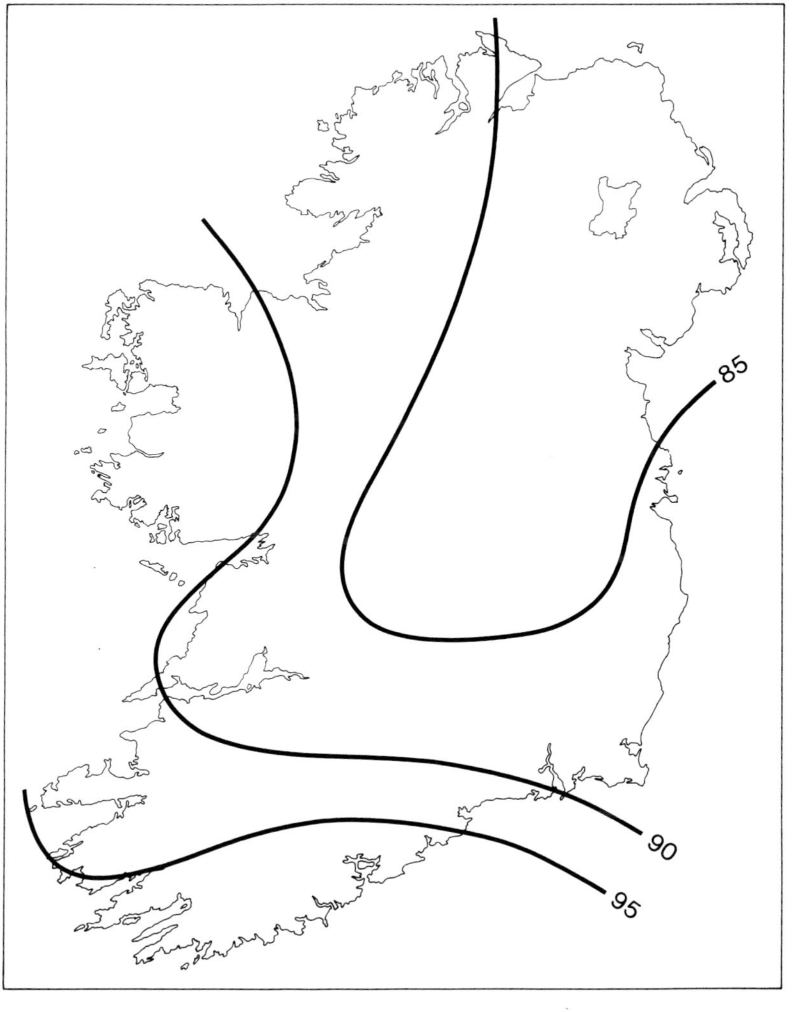

response of the Atlantic Ocean. Winter temperatures can be expected to rise more slowly and less than their summer counterparts and to be accompanied by modest increases in precipitation. Summer precipitation changes are less clear and contradictory evidence is provided by general circulation models at present. On the basis of analogues during the past century, circulation changes would appear to imply that a modest reduction will occur, mostly in eastern and northern areas. Coupled with changes in evapotranspiration this poses potential problems for some aspects of water resource management. The climatic scenario outlined for 2040 in Table 3 suggests a warming in both summer and winter slightly less than that suggested from averaging of five GCM results for north-western Europe by Warrick and Barlow (1991). It does however show many similarities with the working assumptions used by the only national study commissioned on this question to date (McWilliams, 1992), though divergence on the question of summer rainfall changes is apparent. Deciding policy options on the basis of one approach to future climate projection, namely GCM output, may however be risky, and while regional detail may ultimately improve on these approaches, alternatives such as the analogue type approach detailed here should also be used to provide a form of 'ground truth' at the scale of Ireland for climatic simulations which originate from global scale simulations.

Table 3: A Climate Scenario for Ireland for c. 2040

	Temperature	Precipitation
AVERAGE ANNUAL	+2.0°C	Little change
AVERAGE WINTER	+1.5°C	5-10% increase
AVERAGE SUMMER	+2.5°C	5-15% decrease

References

Briffa, K.R., Jones, P.D. and Kelly, P.M., 1990. Principal component analysis of the Lamb catalogue of daily weather types: part 2, seasonal frequencies and update to 1987, *International Journal of Climatology* 10, 549-63.

Crowe, P.R., 1971. *Concepts in Climatology*, London, Longman.

Gates, W.L., Mitchell, J.B., Boer, G.J., Cubasch, U., and Meleshko, V.P., 1992. Climate Modelling, Climate Prediction and Model Validation. In Houghton, J.T., Jenkins, G.J., and Ephraums, J.J., (Eds.), *Climate Change 1992: the Supplementary Report to the IPCC Scientific Assessment*. World Meteorological Organisation/United Nations Environment Programme, Cambridge, Cambridge University Press.

Hansen, J., Fung, I., Lacis, A., Rind, D., Lebedeff, S., Ruedy, R. and Russell, G., 1988. Global climate changes as forecast by Goddard Institute for Space Studies three dimensional model, *Journal of Geophysical Research* 93, 9341-64.

Henderson-Sellers, A. and McGuffie, K., 1987. *A Climate Modelling Primer*, Chichester, Wiley.

Karoly, D.J., 1987. Southern hemisphere temperature trends: a possible greenhouse gas effect, *Geophysical Research Letters* 14, 1139-41.

Lamb, H.H., 1950. Types and spells of weather around the year in the British Isles: annual trends, seasonal structure of the year, singularities, *Quarterly Journal of the Royal Meteorological Society* 76, 393-429.

Lamb, H.H., 1972. British Isles weather types and a register of the daily sequence of circulation patterns, 1861-1971, *Geophysical Memoir* No.116, H.M.S.O., London, Meteorological Office.

Leggett, J., Pepper, W.J. and Swart, R.J., 1992. Emission scenarios for the IPCC: an update, In Houghton, J.T., Jenkins, G.J., and Ephraums, J.J. (Eds.), *Climate Change 1992: the Supplementary Report to the IPCC Scientific Assessment*. World Meteorological Organisation/United Nations Environment Programme, Cambridge, Cambridge University Press.

Lockwood, J.G., 1993. Impact of global warming on evapotranspiration, *Weather* 48(9), 291-298.

Manley, G., 1974. Central England Temperatures: monthly means 1659 to 1973, *Quarterly Journal of the Royal Meteorological Society* 100, 389-405.

Mayes, J.C., 1991. Recent trends in summer rainfall in the United Kingdom, *Weather* 46(7), 190-5.

McWilliams, B., 1991. Executive Summary, In, McWilliams, B. (Ed.), *Climatic Change: Studies on the implications for Ireland*, Dublin, Department of the Environment, 125-171.

O'Hare, G., and Sweeney, J., 1992. Lamb's circulation types and British weather: an evaluation, *Geography* 78(1), 43-62.

Rowntree, P.R., 1990. Estimate of future climate change over Britain, *Weather* 45(3), 79-88.

Stouffer, R.J., Manabe, S. and Bryan, K., 1989. Interhemispheric asymmetry in climatic response to a gradual increase of atmospheric CO_2, *Nature* 342, 660-2.

Sweeney, J., 1985. The changing synoptic climatology of Irish rainfall, *Transactions of the Institute of British Geographers* 10(4), 467-80.

Warrick, R.A. and Barrow, E.M., 1991. Climate change scenarios for the UK, *Transactions of the Institute of British Geographers* 16(4), 387-99.

Washington, W.M. and Meehl, G.A., 1989. Climate sensitivity due to increased CO_2: experiments with a coupled atmosphere and ocean general circulation model, *Climate Dynamics* 4, 1-38.

CLIMATE CHANGE AND ECONOMIC POLICY

John D Fitz Gerald
The Economic and Social Research Institute
4 Burlington Road, Dublin 4, Ireland

1. Introduction

In this paper I want to examine the response by the Irish Government and the European Union to the potential problem of climate change. I consider how recent economic research contributes to our understanding of how this problem can best be tackled.

The problem of climate change and the impact of human economic activity on the atmosphere is rather different from the other major problems which we face in Ireland:

- It is not amenable to solution by individuals acting on their own nor can it be resolved by any independent action of the Irish Government or of any other government. For policy to be effective it must be developed and implemented at a supranational level. The international dimension of the problem is unusual and it makes it difficult for governments throughout the world to find solutions.
- The scientific evidence as to the magnitude of the problem is, as yet, inconclusive. As this conference has demonstrated, there remains considerable doubt about the urgency and magnitude of the danger of climate change arising from human activity. This poses particular difficulty for policy makers. If governments are to face their electorates with unpalatable policies it is not easy if the governments' advisors are clearly uncertain about the necessity or efficacy of their chosen approach.
- Even if the climate is changing, depending on the magnitude of the change, the effects on human welfare may not be very adverse. Analysis for the United States suggests that the economic costs of a moderate increase in temperature could be quite small in the long-term (Nordhaus, 1991). However, because of their greater dependence on agriculture, global warming could prove more serious for third world countries. To date no one has had the temerity to examine how the smaller EU members might fare under global warming. However, the finding of pollen from olive trees on the Céide Fields archaeological site in Mayo could whet Irish appetites for climate change!
- Action to deal with global warming takes the political process into an area of policy discussion with which neither governments nor their electorates are very familiar. Governments every day have to make choices which affect the distribution of income among their current citizens. It is not

often that they have to make such choices explicitly between the standard of living of their current citizens and their citizens' grandchildren.

- Finally, evidence does indicate that, as with super-tankers, the safe braking distance is very long in the case of climatic change. It may be many decades before action to slow the vehicle, in this case greenhouse gas emissions, eventually slows the warming process. This means that failure to act could have a high price in fifty to a hundred years time while action, which may prove to have been unnecessary in fifty years time, will have more certain and immediate economic costs.

While the possibility, or even the probability that economic growth and development may lead to substantial changes in world climate has attracted the temporary attention of world governments at the RIO summit, it is clearly not considered a major or urgent matter for action at a national or an international level. In the case of Ireland, a recent survey undertaken by the Economic and Social Research Institute indicates that households do not see the Greenhouse Effect as being a major environmental problem. Only just over 3 per cent of respondents mentioned it in their top three environmental concerns. Much greater attention was given to other problems such as smog than to the Greenhouse Effect. It is against this background of relative indifference by the electorates of the developed world that governments have to develop policy on climate change. If the nature of the problem could be clearly identified scientifically it might be easier to develop public policies to modify our behaviour, in particular our use of energy.

2. Current Policy

The Irish Government has expressed its concern about the potential dangers of climate change and the Minister for the Environment has said that it is his intention that Ireland, with our EU partners, should ratify the UN convention on climate change as soon as possible. With this in mind the Department published its proposals for limiting carbon emissions in the year 2000 to 120 per cent of their 1990 level (Department of the Environment, 1993). These proposals have been discussed in some detail already at this conference; in this paper I consider how appropriate they are to the task in hand.

These proposals cover the major areas of energy and transport. In the case of energy they include the demand side management programme of the ESB and improved building standards. There is considerable emphasis put on the shift in consumption patterns towards gas. In the case of transport they cover investment in public transport, in particular the Dublin Transport Initiative. In both cases the Department emphasises that the existing fiscal regime tends to tax energy relatively heavily. However, they do not show any enthusiasm for the proposed EU carbon tax and they do not discuss how existing tax and subsidy policy favours solid fuel, including peat, over oil, in spite of the fact that the burning of solid fuel generally results in higher emissions of carbon dioxide than in the case of oil.

The range of measures proposed looks rather ineffective and, realistically, the Department does not claim that they will greatly reduce carbon emissions. They rely more on advice and encouragement than carrots and sticks. If Ireland manages to limit its emissions to the objective the government has accepted it is more likely to be an accident than the result of active policy.

In the case of the demand side measures and the improved building standards, there is an attempt to deal with the problem of energy waste by households. Scott (1993) has shown that households appear to be contrary in the economic sense. They fail to invest in energy saving in spite of the clear economic benefit to them of so doing. However, given this seeming irrationality by households, it seems unlikely that the current measures will prove much more effective for households in existing buildings. It would appear that a more comprehensive programme is needed to help households reap the potential benefits from certain inexpensive energy saving measures.

The policy on public transport is admirable. However, there seems to be little willingness to adopt the logical accompanying measures to force motorists out of cities and to encourage commuters to use public transport. It is interesting to note that the EU Court of Auditors have been critical of the Greek Government who obtained structural funds to build an underground in Athens to reduce air pollution while failing to implement the necessary measures to reduce traffic. It is to be hoped that we will not find ourselves subject to the same criticism in years to come.

There seems to be a lack of enthusiasm on the part of the Irish Government for the EU Commission proposal to implement a carbon tax. Thus far the Irish Government, along with the other peripheral EU countries, has emphasised the costs to Ireland from such a measure. However, as discussed later this approach seems short sighted.

With the exception of the carbon tax, the approach taken by the Irish Government to the problem of global warming is broadly in line with that of the EU Commission. The EU strategy assumes that half the reduction needed to stabilise emissions will come from the carbon tax and the other half will arise from a range of measures, including improved programmes to promote energy saving.

The EU envisages exempting energy-intensive industries from the carbon tax, especially if the tax is not adopted by other major trading blocks. However, in doing so it exempts some of the key industries which might be expected to change behaviour in the face of a tax. The alternative is that these industries might shift location to regions of the world which do not adopt such a tax. Either way, the absence of a concerted policy across the major world trading blocks is likely to reduce the effectiveness of a carbon tax.

3. Economic Growth and Carbon Emissions

Forecasting energy consumption and, as a result, carbon emissions, is very difficult. Among the factors making forecasting difficult is the fact that the economy reacts slowly to changing energy prices. When energy prices rose in the first OPEC oil crisis, energy demand took many years to react fully. This slow response to changing prices reflected the fact that, in many cases, increased energy efficiency is achieved through installing new machinery - a time consuming process - rather than changing the mode of operating existing plant. In addition, when oil prices rose it made research into increased energy efficiency much more profitable than in the 1960s. However, the benefits of this research took many years to bear fruit.

The delayed action response to price changes makes their quantification difficult. Conniffe (1993) argues that there is a ratchet effect in the response to changing prices. When the price of energy rises it encourages research which eventually leads to increases in efficiency. However, once the new energy-efficient technology is developed, even if prices fall, the technology is not *forgotten* and energy efficiency has been raised to a new plateau. When this ratchet effect is taken into account it appears that energy demand in Ireland is still rising in line with increases in GNP, even though simplistic models suggested some slowdown (Conniffe, 1993).

In a paper by Fitz Gerald and McCoy (1992), the ESRI's *HERMES*[1] model was used to prepare a forecast for the Irish economy out to the end of the decade. On the basis of past behaviour and current trends in substituting gas for other fuels, it was estimated that carbon emissions would grow by around 24 per cent over the decade. This growth would take Ireland above the target agreed between the Irish Government and the EU Commission. If growth in output were to prove greater than forecast over the decade then the excess emissions would be even larger. In addition, the medium-term forecast was prepared on the basis of rising real energy prices; to date real energy prices have fallen, encouraging even greater energy consumption than the forecast assumed.

This forecast would suggest that significant policy changes will be required if the target for carbon emissions in 2000 is to be met. It is unlikely that the current set of instruments will achieve their hoped-for result.

4. Effective Policy?

The design of instruments for efficiently implementing environmental policy is easiest where the location or concentration of pollution is not a concern of policy (McKay, Pearson, and Smith, 1990). For example, the use of simple national taxes to control pollution of watercourses would be wholly inappropriate; the same effect can be arrived at much more efficiently by legal remedies. In the case of global warming, where emissions, wherever they occur on the globe, have an equal effect on the environment, taxes are the most efficient instrument to limit emissions. (However, while taxes may be most efficient, they may prove impossible to impose at a global level and some other approach may be required.)

When dealing with many polluters and many individuals who suffer from the pollution, some form of market-based instruments, such as taxes, should be used to ensure an efficient allocation of the burdens of environmental policy (Scott, 1992). Recourse to regulation, while apparently simple, is likely to have high costs, due to the imperfect information available to the regulatory authority (Hahn, 1993) and due to the problems of implementation and policing.

The disadvantage of a regulatory approach to the global warming problem, based on national or individual quotas, is that the choice of quota is essentially arbitrary. The same quota per head or per unit of GNP may impose very different costs on different countries or individuals. For example, carbon dioxide quotas (based on population or GNP) might impose no costs on a country with ample supplies of hydro-electricity but it could impose very heavy costs on another country which had a different structure of natural resources or of productive capacity. Similarly, the same quota applied to individual households or industries could impose very different costs. To choose quotas which will minimise the cost to society of limiting carbon emissions will require very detailed information on the characteristics of all those involved, an impossible task for any government.

Every restriction on production or consumption, if it is to have any effect, will impose a cost. This cost will have an identical effect on the individual and on the economy as the imposition of a tax of the same amount as the cost. The difference lies in the fact that the cost (tax) will vary from individual to individual and from economy to economy. These differences will reflect the magnitude of the problems caused by the regulation for the different individuals. By contrast, a tax imposes an equal cost on all those who have to pay it.

While there is definitely a need for other types of policy, especially where households or firms are apparently acting irrationally (Scott, 1993), the most appropriate method for achieving a stabilisation of carbon emissions in the long term is a carbon tax.

Over the last three years many studies have been carried out to examine the potential economic effects of carbon taxes. These studies have considered the likely impact on the world as a whole, on the United States, on the European Union and on individual member states, including Ireland. The methodology used to examine the effects has included the use of macro-econometric models, such as the HERMES model of the EU and the ESRI's Medium Term Model, and general equilibrium models such as the OECD's *GREEN* model covering the world as a whole, and similar models for the United States.

While these studies have produced rather different estimates of the economic effects of carbon taxes there appear to be a number of conclusions which are confirmed by a range of different studies.

Firstly, whatever about their economic effects on individual countries, any economic measures to limit the growth of carbon emissions would have to be implemented throughout the world. It would not be possible for any one major country, or even a single major trading group such as the EU, acting on its own, to halt the rise in emissions. Unless all major trading groups and the major economies of the Third World adopt a concerted policy it will be a failure (Burniaux, Martin *et al.*, 1992). The sheer size of economies such as China means that they cannot be ignored. If they are, the restrictions or taxes imposed in other countries will result in industry which is energy-intensive moving location to the regions or countries which fail to act. The result will be little change in emissions but significant costs for those countries where energy intensive industries are penalised.

The second broad conclusion is that even tax changes on the scale of those proposed by the EU will not be sufficient to halt the rise in carbon emissions. Burniaux, Martin, *et al.* (1992), using their GREEN model, suggest that a tax of between \$150 and \$200 a ton of carbon might be needed by the year 2020 to stabilise emissions. For Ireland the \$20 a tonne tax suggested by the EU would only achieve a small reduction of under 3% in Irish emissions by the year 2000 (Fitz Gerald and McCoy, 1992).

The third broad conclusion from the range of studies now available is that for tax changes of the order currently proposed by the EU Commission, the macro-economic effects are unlikely to be very significant. Even for the much larger tax changes examined by OECD with their GREEN model, the costs are estimated to be of the order of 2 per cent over thirty years, a relatively small fraction of the economic growth which could be expected over such a period. In estimating the results of the taxes, account must always be taken of the effects of spending the revenue from the tax: either using it to reduce other taxes or to increase public expenditure. Whether or not the macro-economic effects of a carbon tax will be detrimental to an economy cannot be determined *a priori*. It will depend on the size of the tax, how the receipts are used, and the structure of the economy.

One study for the major economies of the EU (Standaert, 1992) suggests that for a $10 tax the long-run effect would be a reduction in GNP of between 0.1 per cent and 0.5 per cent. Bradley (1992) argues that the specification used by Standaert in the HERMES model is unsatisfactory and that the resulting simulations exaggerate the costs. The model does not allow for the fact that wage rates may respond to changes in taxation.[2]

When account is taken of this mispecification it is quite likely that within the EU a carbon tax, combined with a reduction in social insurance contributions, would actually increase welfare *before* taking account of the environmental benefits. It would also be likely to lead to an increase in employment. This potential benefit arises from the distortion of prices in the economy arising from high rates of labour tax in most EU countries.

A study for Ireland using the Irish version of the HERMES model[3] (Fitz Gerald and McCoy, 1992) suggests that the $20 tax proposed by the EU Commission, if combined with a reduction in social insurance contributions, would actually increase GNP in Ireland by around 0.4 per cent. This arises from the fact that the tax on labour has a significant effect on employment whereas the tax on carbon will not, as discussed above, lead to a major change in behaviour.

The fourth broad conclusion from these studies is that countries, regions, or companies which earn a significant part of their income from energy production would be likely to suffer most from a carbon tax. In the case of a global tax OECD suggest that the OPEC oil producers would suffer a very significant loss of income as a substantial part of the incidence of the tax would fall on them in the long term. The reduction in GNP in energy exporting LDCs could be up to five times that experienced by Europe.

A recent study for the United States suggests that a carbon tax would have a small negative effect on GNP (Goulder, 1992). The substantial negative effects of the tax on energy producers and oil refiners would more than offset the benefits from reductions in other distortionary taxes.

For Europe as a whole and for Ireland in particular, the limited importance of the energy sector and energy production means that a carbon tax, combined with a reduction in other distortionary taxes, is quite likely to have a positive effect on economic welfare through raising GNP and employment. However, individual industries may be badly hit where they are particularly dependent on energy production. In addition, the income distribution consequences of a tax are likely to be negative in some countries including Ireland (Scott, 1992), the United States (Poterba, 1992), and the United Kingdom (Johnson, McKay and Smith, 1990), though it may have little effect on the distribution of income in certain other European countries (Smith, 1992).

The fifth broad conclusion is that part of the reason for the small negative (or even positive) effects of a carbon tax along the lines proposed by the EU Commission is the fact that energy users will not greatly alter their behaviour in response to the tax. From the environmental point of view this is obviously a disadvantage as it means a very large tax will be needed to achieve a stabilisation in carbon emissions.

All studies of energy demand indicate that the initial reaction to a change in prices, in this case induced by a carbon tax, is small. However, as discussed above, over a long period a rise in taxes encourages new energy-saving investment and research into alternative technologies (Conniffe, 1993). Because the benefits of research take considerable time to manifest themselves the proposals for a carbon tax envisage a gradual escalation of its magnitude up to the end of the decade. This time scale would encourage research and investment in energy-saving and alternative non-fossil energy sources and it would allow time for the economy to adjust. The costs of controlling carbon emissions are likely to be greatly reduced if advanced warning is given of the controls or taxes.

Finally, all the studies, in indicating that very high taxes would be needed to achieve a substantial reduction in carbon emissions below current levels, show how difficult it will be to achieve such a change. The fact that high taxes would be needed should not be taken to mean that other methods, such as regulation and control, are likely to be cheaper and pose less of a burden on the economy. In fact, as discussed above, regulation and control would almost certainly prove much more of a burden for the economy and the affected sectors.

Conclusions

Policy makers, perforce, must decide their approach to the potential problem of global warming without the benefit of clear-cut scientific evidence. Global warming might not happen or if it does it might not have serious consequences. However, there remains the distinct possibility that it could prove very serious for all those living on the planet at some date over the next few centuries. Faced with this prospect it seems wise to take action to avoid the possibility that the pessimists are right.

However, given the uncertainty and the time scale of the problem it seems right to take action gradually. We have time to experiment to find the least costly manner of achieving the environmental objective of slowing, and possibly reversing carbon emissions. The costs will be minimised if governments, including the Irish Government, signal the potential rise in the cost of carbon emissions well in advance. This will encourage research into energy-saving and alternative non-fossil fuel sources.

In Ireland it would appear that we have the possibility of obtaining a *double dividend*. It seems likely that a carbon tax of the magnitude proposed by the EU Commission would leave the Irish economy better off than to-day with more people employed. These benefits would arise because the application of the revenue from the carbon tax to reducing taxes on labour would encourage employment.

Both at the level of the European Union and in individual economies, such as Ireland, many existing policies serve to distort the market, encouraging carbon emissions and discouraging research into certain alternative energy sources. In the EU, the CAP massively raises the price of land by encouraging farmers to produce unwanted output. In a free market land prices would be much lower and the economics of biomass would be more attractive.

In may EU countries, including Ireland, there is a policy of subsidising domestic production of solid fuel. The effect of this policy is to increase carbon emissions above the level they would otherwise be, reduce the efficiency of the economy, and give the wrong signals for research.

It seems time for governments within the EU to examine the coherence of their existing policies and to consider whether a change in these policies could contribute more to reducing carbon emissions than the new measures currently under consideration.

Finally, if there is to be any possibility of success for attempts to control carbon emissions and to halt the process of global warming, new policies will have to be adopted at an international level by all major economies and trading groups. While we in Ireland can take 'a principled stand' on our own, possibly at little or no short-term cost, unless we, through the EU, persuade other major economic groups to do the same there will be no effect on global warming.

Footnotes

1 This model is now known as the ESRI's Medium Term Model. In this paper the model was augmented with an extended energy sector to analyse the effects of carbon taxes.

2 Drèze and Bean (1990) show that taxes can affect wage rates in some of the countries examined.

3 Now the 'Medium Term Model'.

References

Burniaux, J., Martin, J., Nicoletti, G., and Oliveira Martins, J., 1992. The Economic Costs of International Agreements to Reduce Global Emissions of CO_2, InFitz Gerald, J., and McCoy, D. (Eds.), *The Economic Effects of Carbon Taxes*, Policy Paper No. 14, The Economic and Social Research Institute.

Conniffe, D., 1993. Energy Elasticity Estimates and the Stability of the Relationship with GDP, in Fitz Gerald, J., and McCoy, D. (Eds.), *op. cit.*

Department of the Environment, 1993. *Ireland: Climate Change - CO_2 Abatement Strategy*, Dublin, Stationery Office.

Dr0ze, J. H., and Bean, C. R., 1990. *Europe's Unemployment Problem*, The MIT Press, Cambridge Massachusetts and London, England.

Fitz Gerald, J., and D. McCoy, 1992. The Macro-Economic Implications for Ireland of Carbon Taxes, In Fitz Gerald, J. and D. McCoy (Eds.), *op. cit.*

Goulder, L. H., 1992. Carbon Tax Design and US Industry Performance, In Poterba, J. M. (Ed.), *Tax Policy and the Economy*, MIT Press, Cambridge Massachusetts.

Hahn, R. W., 1993. Comparing Environmental Markets with Standards, *Canadian Journal of Economics*, XXVI, No. 2, May.

Johnson, P., McKay, S., and Smith, S., 1990. *The Distributional Consequences of Environmental Taxes*, London, The Institute for Fiscal Studies.

McKay, S., Pearson, M., and Smith, S., 1990. Fiscal Instruments in Environmental Policy, *Fiscal Studies*, November.

Nordhaus, W. D., 1991. Economic Approaches to Greenhouse Warming, In Dornbusch, R., and Poterba, J. (Eds.), *Global Warming: Economic Policy Responses*, London, MIT Press.

Pearson, M., and Smith, S., 1991. *The European Carbon Tax: An Assessment of the European Commission's Proposals*, London, The Institute for Fiscal Studies,

Poterba, J. M., 1991. Tax Policy to Combat Global Warming: On Designing a Carbon Tax, In Dornbusch, R., and Poterba, J. (Eds.), *op. cit.*

Scott, S., 1992. Carbon Taxes: Theoretical Considerations and Estimates of the Effects on Households, In Fitz Gerald, J., and McCoy, D. (Eds.), *op. cit.*

Scott, S., 1993. Energy Conservation in the Home - Are We Contrary?, In Fitz Gerald, J., and McCoy, D. (Eds.), *Issues in Irish Energy Policy*, Policy Paper No. 20, The Economic and Social Research Institute.

Smith, S., 1992. Public Finance Aspects of a European Carbon Tax, *European Economy*, Special Edition No. 1.

Standaert, S., 1992. The Macro-Sectoral Effects of an EC-Wide Energy Tax, In *European Economy*, Special Edition No. 1.

THE ROLE OF THE IRISH METEOROLOGICAL SERVICE IN CLIMATE CHANGE

Declan Murphy
Meteorological Service

1. Introduction

The Meteorological Service is the national body charged with responsibility for operational meteorology and climatology. As might be expected, its main day to day concerns are with weather forecasting and the routine of building up our store of climatological data. Nevertheless the Meteorological Service does have a role to play in the more long-term issue of climate change. While not at the forefront of climate modelling it is involved in international projects, and sees its archive of Irish climatological data as a valuable resource for Irish and international researchers.

Given the fact that the composition of the atmosphere is changing, it is logical to assume that the world's climate might be affected. There is no room for complacency, and the uncertainties which attach to the prediction of climate change are no excuse for inaction. There is a need to try to understand what is going on, to assist in the task of predicting change, to evaluate possible impacts and to take steps to cope with what the future might hold and, indeed, to exploit opportunities which climate change might offer. There is also a responsibility to inform the public and the decision-makers. A wide variety of scientific disciplines is required to tackle the questions posed by the climate change issue, as well as economic, political and communications expertise. We must ensure that a way is found to bring all the relevant knowledge and skills to bear on the subject and to build structures that will make that happen.

2. The Role of the Meteorological Service

The areas relating to climate change in which the Meteorological Service sees itself having a responsibility are :

- the development and maintenance of an archive of Irish climatological data sufficient to define the climate of the country, to detect variation in Irish climate and to be available as a resource for climate and climate change research;
- international co-operation with a view to building up regional and global climate datasets;
- monitoring the results of climate prediction studies, on a global and regional scale;
- stimulating and supporting research on Irish climate, past, current and future;

- provision of advice to the government and to the public on the state of Irish climate and on future trends;
- contributing to research on impacts of climate change in Ireland;
- contributing as appropriate to the development of national response strategies.

3. Climate Datasets

The Meteorological Service devotes considerable resources to development of a comprehensive, quality-controlled archive of Irish climatological data, drawing in data from its own stations and from a variety of voluntary and co-operating individuals and organisations. This archive is described elsewhere in this volume.

The data are made available to any research project in Ireland or abroad. In general the climate archive is seen as part of the fundamental scientific dataset which any country needs to develop for current and future needs.

The Service also has become involved in international efforts to produce coherent datasets on a regional and global basis for use in the detection of climate change and in climate modelling. Examples are the North Atlantic Climate Dataset (co-ordinated by Denmark) and the 'Re-analysis' project organised by the European Climate Support Network and carried out at the European Centre for Medium-range Weather Forecasts, with the aim of producing a fifteen-year set of data in gridpoint form for input into climate modelling.

Some research into Irish climate is done within the organisation and the Service would see its role as including the encouragement and support of other relevant efforts in the country.

4. Climate Prediction

The Meteorological Service monitors developments in the prediction of climate change, principally through the findings of the IPCC and through other initiatives with which the World Meteorological Organisation is involved. It has a particular interest in regional studies of our area of interest. It must be admitted that our involvement in this area, apart from international programmes which we help fund as part of our membership of international organisations, has largely been that of receiver and interpreter of other people's work. While climate modelling is too ambitious for us, a more pro-active approach to regional climate studies and to the application of the results of large-scale models would be desirable and beneficial.

International involvement includes the recent development of the European Climate Support Network, set up by the meteorological services of western

Europe to achieve better co-ordination of climate research activities in major centres such as the Hadley Centre in England and the Max Planck Institute in Germany, and to engage in joint programmes of benefit to all, such as compilation of climate datasets and model intercomparison studies.

5. Advice on Climate Change

Through whatever means are available to it, the Meteorological Service takes the opportunity to inform the public on the subject of climate change. Quite often it finds itself in a position of having to correct and deflate reports or opinions carried in newspapers, usually quoting or misquoting some researcher in the United States on the subject, either of an extreme alarmist nature or debunking the notion of climate change entirely. As is often the case with the media, a sober evaluation of scientific data is not as quotable or as newsworthy as an individual riding a hobby horse.

The Service also has a role in advising the government on matters relating to scientific aspects of climate change. Since the issue is essentially an 'environmental' one the Minister for the Environment is obviously the Cabinet member most likely to deal with matters related to it. The fact that the Meteorological Service is attached to a different Department, i.e. Transport, Energy and Communications, means that the linkage with the Minister and his officials is not as strong as might otherwise be the case. Some ideas on improving the structures are given below.

6. Impacts and Response Strategies

Although the Service's expertise lies mainly in the science of climatology, it has a part to play in studies of the impacts of climate change on various economic sectors. The provision of climatological datasets and of advice on trends in temperature, rainfall etc. is obviously the main way in which the Service can help, but its expertise in certain applications of meteorology (such as agricultural meteorology) could be of great value in an interdisciplinary approach.

In the development of national strategies in response to the challenges of the climate change issue the Meteorological Service can also assist in a manner appropriate to its role as a national scientific organisation.

7. Structures Necessary for a National Response

There is a variety of organisations which are in some way active in the area or who feel some responsibility for it in the country. The role of the Meteorological Service, both present and potential, has been spelt out. The Department of the Environment has a statutory role in the area, and other government departments also have an interest arising from the economic sector they regulate. State bodies such as ESB and Teagasc also have an interest. There are active research programmes in related areas within the universities.

There is, however, a lack of co-ordination of these interests and activities. There is no structure to bring together the interested parties or to develop an overview of the issue as it pertains to Ireland. To a large extent the response of some state bodies up to now has been reactive . The Meteorological Service has been conscious for some time of the desirability of creating some kind of national forum which might address this need. A Climate Study Group set up by the National Committee for Geodesy and Geophysics has considered what might be done, and has promoted the idea of a National Climate Committee on the lines advocated by the World Meteorological Organisation. Basically the WMO concept envisages that the work of the IPCC would be reflected at national level, with attention being given to climate data and prediction, to impacts and to response strategies. An *ad hoc* committee arising from the Study Group is now in the process of formulating a proposal for a National Committee on Climate Change and Ozone Depletion. (It was decided to combine it with the issue of stratospheric ozone depletion as there are scientific connections and because the two problems are inextricably linked in the minds of the public and the media). The basic aims of the proposed committee would be as follows:-

- to stimulate research into the scientific aspects of climate change and the impacts of climate change;
- to provide an authoritative voice on the subject within the country;
- to provide input into the development of response strategies;
- to inform the public and the media on the issue.

The composition of the Committee would be drawn from government Departments, the Meteorological Service and other state bodies, from semi-state bodies and from the private sector. Some kind of sub-structure would be required to address specific aspects of the question. The reporting relationship would be to the Minister of the Environment.

It would also be the task of the Committee to make a case for resources to be made available for work in the whole area, and to influence the allocation of government funding. Financial and human resources would also be necessary for the operation of the committee itself. The organisations which might be candidates for hosting it are fully stretched already, but it is hoped that a solution will be found.

References

Frich, P. and Cappelen, J., (Eds.), 1992. Homogeneity testing and correction of spatial and temporal inhomogeneities in North Atlantic climatological data series, *Proceedings from the Third Nordic Climate Workshop*, Technical Report 92-15, Danish Meteorological Institute, 2 - 13.

European Climate Support Network, 1993. *Outline of the National Meteorological Services European Climate Support Network*, 17 - 23.

CLIMATE CHANGE: CURRENT EU POLICY AND THE IRISH CONTRIBUTION

Mary Moylan
Department of the Environment

EU policy on climate change has developed from principles set down in the Declaration on the Environment adopted at the European Council in Dublin in June 1990. The Declaration noted the special responsibility of the Community and its Member States to encourage and participate in international action to combat global environment problems.

In line with this, a joint meeting of Environment and Energy Ministers declared in October 1990 the Community's near-term objective as stabilisation of CO_2 emissions in general by the year 2000 at 1990 levels in the Community as a whole. It was recognised that countries with, as yet, relatively low energy requirements which could be expected to grow in step with their development, would need targets and strategies which would accommodate that development while improving the energy efficiency of their economic activities. The Council also pointed to the potential of world forests to act as a sink for greenhouse gases and, in this regard, advocated the initiation of vigorous programmes of forest protection and development.

UN Convention

The United Nations Framework Convention on Climate Change was signed by over 150 States in June 1992 at the 'Earth Summit' in Rio de Janeiro. The ultimate objective of the Convention is the 'stabilisation of greenhouse gas concentrations in the atmosphere at a level that would prevent dangerous anthropogenic interference with the climate system'. The Convention provides that developed countries should take the lead in combating climate change 'on the basis of equity and in accordance with their common but differentiated responsibilities and respective capabilities' .

The Convention enables regional economic organisations such as the EU to implement the Convention jointly. The European Community has ratified the Convention on behalf of the Union and Ireland will ratify shortly. The Convention will come into force at the end of March 1994.

EU Policy

As stated earlier, the EU policy is to stabilise Community emissions of CO_2 at 1990 levels by the year 2000. This policy will be implemented by a combination of measures at Community and national level.

At Community level, agreement has already been reached on measures on energy efficiency and the use of renewable fuels under the SAVE and ALTENER Programmes. A Directive on energy labeling of domestic appliances has already been adopted under the SAVE Programme and the Department of Transport, Energy and Communications will be launching an information exchange programme later this year.

Agreement has also been reached on a mechanism to monitor implementation of climate change mitigation policies in Member States. Under this Decision, Member States will be required to submit inventories of CO_2 emissions and details of national policies. This will enable the Commission to assess the progress being made towards achievement of the Community's stabilisation policy.

The European Commission has also proposed a carbon energy tax as part of Community policy on mitigating the effects of climate change. The proposal is a complex one, and has serious policy implications for all Member States because of its impact on energy pricing, industrial competitiveness and national budgetary discretion. The Commission proposal is that the tax would be levied 50% on the energy content of the fuel and 50% on the carbon content. The tax would be introduced at the rate of $3 per barrel of oil equivalent and rise at the rate of $1 a year to $10 a barrel.

The government has considered the proposal carefully and is not opposed provided certain concerns of ours are addressed. These concerns relate to

* the need for a provision to allow the EC Cohesion States to modulate the proposals to their relative advantage;
* a satisfactory provision as regards the impact of the tax on the peat sector;
* an effective conditionality clause so as to maintain the competitiveness of EU industry generally;
* an effective provision to maintain the competitiveness of energy-intensive firms based within the European Union.

The negotiations have been very long and detailed and considerable work remains to be done before agreement can be reached. At the moment we are awaiting the views of the new Greek Presidency on how the proposal should be progressed.

National Measures

National plans are an essential component of the EU strategy. Ireland submitted its national strategy to the European Commission and the Strategy was published in June last year. Like the Climate Change Convention, EU policy recognises that a number of Member States, including Ireland, need targets and measures which will accommodate necessary economic growth. Increased energy demand from economic expansion, as well as other structural factors such as reliance on peat, a high carbon fuel, for about 15% of our energy needs, our already high use of low-carbon natural gas, and the absence of a nuclear energy option, affect Ireland's ability to reduce CO_2 emissions.

Our national strategy is based on the objective of limiting national CO_2 emissions to a maximum of 10.7 million tonnes of carbon by the year 2000, an increase of 20% over the 1990 level. If account is taken of the increased carbon sink capacity from the national afforestation policy, the actual growth is 11%. Since a continuation of existing policies would indicate an increase greater than this, achievement of the 20% target will require a co-operative effort on the part of all the different sectors involved: energy producers, industrial and commercial consumers and private citizens.

The National Strategy contains a programme of measures in the areas of energy conservation, fuel use, transport, waste management and afforestation. These measures are designed to limit the levels of carbon in the atmosphere and to improve the energy efficiency of our economy.

Energy Sector

A number of programmes are under way including the following:

* The ESB is pursuing an active Demand Side Management Programme. The intention is to manage load growth around the level of 3% per annum while still catering for national economic expansion.
* New insulation standards for new buildings incorporated in the Building Regulations 1991 are expected to reduce CO_2 emissions from this source by 2% by the year 2000.
* Fuel switching favourable to CO_2 reduction will be promoted by the continued extension of the natural gas grid in the residential and industrial sectors, backed up by the new pipeline from the UK.

Transport

Transport is a significant source of CO_2 emissions, and the largest concentration of transport is in the Greater Dublin area. The principal objective will be to improve public transport and reduce traffic congestion. The report of the Dublin Transport Initiative will be published shortly, and in developing its strategy full account has been taken of environmental impact factors. The new high specification energy-efficient buses introduced by Dublin Bus will compete with the car in Dublin city and the commuter rail services will be further developed.

Because of the dispersed nature of Ireland's rural population, transport needs there will continue to be met generally by the private car, but public transport links will be developed as resources permit.

Waste

The decay of waste containing carbon results in emissions of methane to the atmosphere and to a lesser extent of CO_2. The Department is developing a national waste strategy and in the context of the climate change strategy any reduction in the volume of waste for final disposal is of great importance.

Afforestation

Green plants act as a sink or trap for CO_2, thereby reducing the CO_2 content of the atmosphere. As Ireland is the least forested area within the EU, it is clear that greater afforestation has the potential to make a significant and cost effective contribution to our climate change strategy. The government's annual target for planting is 30,000 hectares up to the year 2000. This is expected to increase CO_2 absorption capacity by 0.8 million tonnes of carbon by the end of the decade. This will provide a substantial counter-balance to the expected increase in carbon emissions over the same period.

Implementation

An inter-Departmental group has been set up to monitor implementation of the national strategy. The principal task facing the group at the moment is the preparation of inventories of greenhouse gas emissions. The Inter-governmental Panel on Climate Change and the European Commission are working on internationally acceptable methodologies. Considerable work is also under way at the OECD on developing agreed methods for reporting the information required under the Convention. While considerable progress has been made, the reporting obligations are quite onerous. The Environmental Protection Agency, along with the Department of Transport, Energy and Communications and the Department of the Environment, are working on the Irish statistics at present.

It is essential that the strategy be subject to formal review given the many uncertainties which surround the science of climate change. The Strategy will be reviewed in accordance with EU policy and the requirements of the UN Convention. The first meeting of the parties to the Convention is scheduled to take place in April 1995 and the reporting arrangements will be agreed at that stage.

Conclusion

The Department of the Environment recognises that the national climate change strategy poses many challenges. We are confident however that it can be implemented successfully and that it will bring many environmental and economic benefits to the country.

Round Table Discussion

(Chairman: Declan Murphy, Director, Meteorological Service)

Declan Murphy opened the discussion by reminding the participants of the questions posed by John Maunder at the opening of the conference: 'What *is* the Irish response to climate variation and climate change? What *ought* to be our response?' He then asked Mary Moylan of the Department of the Environment (DoE) exactly what sort of input the Department would like from 'the community', i.e. those working in the climate/climate change area.

Mary Moylan said that the Department of the Environment is currently concentrating on implementing the CO_2 Abatement Strategy announced last summer. Inventories of greenhouse gases are being prepared, mainly by the Environmental Protection Agency/Environmental Research Unit, in co-operation with the Department of the Environment and the Department of Transport, Energy and Communications. She said that the Department sees climate data as important in the assessment of the impact of policies. They also see climate modelling as an important activity. The Department of the Environment is involved in the IPCC but not very actively. Finally she said that while the Department of the Environment is very interested in scientific work in this area, it only had limited resources for use in supporting such work. She felt the conference had been very useful as a networking process.

Aodhagán Roddy from UCG reminded the Conference of the importance of not overlooking greenhouse gases other than carbon dioxide - such as methane, which is (i) increasing more rapidly than CO_2, (ii) is more effective in absorbing infra-red radiation, and (iii) emissions of which can be more easily reduced.

Kevin O'Rourke from the Energy Group in Forbairt said that CO_2 abatement is a 'no regret' strategy in that it leads to greater energy efficiency and conservation, which are desirable objectives even if the expected climate change does not come about. He said that financial instruments, such as a carbon tax, and the provision of information and advice were both important means of bringing about reductions in CO_2 emissions but he said that there is also a case for regulation, particularly insofar as building standards are concerned. Both technology and economics favour this approach, as can be seen in the case of Denmark. He said there is a need for a proper balance between these three types of instrument.

John Fitz Gerald from the ESRI accepted the need for regulation in cases of what he termed 'market failure', but he argued that regulation was insignificant on longer time scales, where taxation was a more effective instrument. His main concern, he said, was to focus on the present absence of taxation as a CO_2 abatement instrument.

John Maunder spoke about the proposed National Committee on Climate Change and Ozone Depletion. He described the structure of the World Climate Programme and showed how the various international organisations are involved. He suggested that the structure of the proposed National Committee might follow this model. He said that the experience of other countries' national committees could provide useful guidance for Ireland, and referred to Canada, which he said has a good National Climate Programme, which includes the research community, universities, the private sector (as far as possible, e.g. representation of the Wheat Board), and provincial and federal Government. He also mentioned the difficulties experienced in Australia in setting up a national climate committee arising out of a conflict between the Bureau of Meteorology and the CSIRO.

Declan Murphy then asked if there was a need for more and/or better structures.

Owen Wilson from the ESB asked what additional value the Canadian Committee provided.

John Maunder replied that one advantage of the Canadian Committee was that it was not run by any Government department. The Chairman was independent and had direct access to the Minister.

Stu Daultrey suggested that Irish policy on climate change should be nothing, and further suggested that Ireland should urge the EU to do nothing! He said that impact studies show that climate change, if it happens, will be good for Ireland, in most cases. In principle, he said, we should adapt/respond to environmental change, rather than try to stop it.

Chris Shouldice said that the proposed committee should seek to maximise the advantages of our environment and changes thereto. He favoured a broad-based committee, and suggested that one step the committee could take would be to recommend that the Department of Finance remove VAT from attic insulation.

However **John Fitz Gerald** responded that people don't respond to price changes in this area. Responding to Stu Daultrey, he said that even if global warming was not as serious a problem as first suggested, and the impacts on Ireland might be favourable, the Government should nonetheless not take risks.

However **Kevin Byrne** from the Insurance Corporation of Ireland said that we should not wait for a major event that would cost a great deal; the question is whether or not we take action now to mitigate such a possibility.

Another participant raised the question of the impact of global warming on the Third World. Would it lead to more droughts, more flooding?

Aodhagán Roddy said that even though there was no certainty on the issue, the majority scientific opinion supported the likelihood of global warming and we should therefore act now to reduce this likelihood. He supported the idea of a national committee which should come up with a national consensus on this question and advise Government on what should be done.

Brendan McWilliams said it was true that there is still no certainty about global warming, that the possible impacts on Ireland do appear favourable, and that as climate models become more sophisticated they are tending to reduce the extent of the likely warming. However although it is easy to be complacent at the moment, he said there was still a need for a national climate committee to monitor this whole area on an ongoing basis and to analyse, assess, and ultimately advise the Government.

Declan Murphy asked if there was anything that we in Ireland - meteorologists and climatologists - can do to reduce this uncertainty.

Kevin Byrne said that better and more up-to-date information about potential impacts was required. He pointed out that international re-insurers see climatic changes already occurring, more frequent storms for example. Eventually they will not provide cover. The Irish insurance industry needs to be in a better position to rebut the arguments advanced by these international re-insurers; they need more information that is specific to Ireland, given that the re-insurers tend to consider Ireland in the same climatological context as the UK.

Tom Keane said there was a need to co-ordinate scientific effort in this country. Climate should be seen as a natural resource, with natural variability, which may change over time.

John Sweeney said that perhaps the uncertainty about global warming/climate change was over-emphasised. He said there was a need for further impact studies, and expressed concern at the danger of doing nothing. He also supported the proposed national committee but said there was a need for a clearly defined role and responsibilities, together with a need for a lead from central Government.

John Martin from the Office of Public Works also favoured the establishment of a National Climate Committee. As an engineer he needs information on flood frequencies and magnitudes for bridge design, for example. He noted that an extra 300mm is being added to embankments in Northern Ireland in view of the possibility of climate change. He needs more guidance. He cited the Irish committee of the International Hydrology Programme as a possible model for the National Climate Committee. The Irish committee is composed of representatives of universities, Government Departments and consulting engineers, and it is funded by the OPW. He suggests that the DoE should fund the National Climate Committee.

John Maunder argued that we should speak of a Climate Committee rather than a Climate Change Committee, because climate should be seen as a resource. He pointed out that this conference was focused on the Irish situation. Finally he referred to a letter from Bord Fáilte stating that existing climate change information was not adequate for the purposes of tourism planning.

Frank Convery noted that when people get together the band of error tends to narrow. The proposed national committee would provide the best advice to the Minister and to various interested bodies such as the OPW, Bord Fáilte, the insurance industry etc. The committee could come from 'line agencies' and the universities. He expressed concern about a 'very official' committee. He supported the concept of a national committee and recommended a 'bottom-up' approach. He concluded by thanking John Maunder for organising the conference.

Declan Murphy concluded by identifying certain strands in the discussion.

1. Climate change *may* affect Ireland in benign ways, but Governments could not afford to gamble.
2. There appeared to be support for some sort of committee or group.
3. We must also think about the rest of the world, and not just Ireland.
4. Climate should be seen as a resource, not just as a possible problem.

He brought the discussion to a close by thanking the various speakers, Frank Convery for providing the facilities of the Environmental Institute, and John Maunder for organising the conference.